The publisher gratefully acknowledges the contribution toward the publication of this book provided by the General Endowment Fund of the Associates of the University of California Press.

A Fascination for Fish

UNIVERSITY OF CALIFORNIA PRESS /
MONTEREY BAY AQUARIUM SERIES
IN MARINE CONSERVATION

"Healthy oceans are critical to the future of all life on Earth, yet by and large the underwater world remains hidden to us, unknown and mysterious. The mission of the Monterey Bay Aquarium is to inspire conservation of the oceans, and this series of books is intended to further that goal. By helping people discover their connection with the natural world, we hope to foster a lifelong commitment to learning about and caring for the oceans on which all life depends."

—JULIE PACKARD
Executive Director
Monterey Bay Aquarium

1. *The Monterey Bay Shoreline Guide,* by Jerry Emory
2. *A Living Bay: The Underwater World of Monterey Bay,* by Lovell Langstroth and Libby Langstroth
3. *A Fascination for Fish: Adventures of an Underwater Pioneer,* by David C. Powell

DAVID C. POWELL

A Fascination for Fish

Adventures of an Underwater Pioneer

WITH A FOREWORD BY
Sylvia A. Earle

UNIVERSITY OF CALIFORNIA PRESS
BERKELEY · LOS ANGELES · LONDON

MONTEREY BAY AQUARIUM

University of California Press
Berkeley and Los Angeles, California

University of California Press, Ltd.
London, England

MONTEREY BAY AQUARIUM®

Library of Congress Cataloging-in-Publication Data

Powell, David C., 1927–.
A fascination for fish : adventures of an underwater pioneer / David C. Powell ; with a foreword by Sylvia A. Earle.
p. cm. — (University of California Press/ Monterey Bay Aquarium series in marine conservation ; 3)
Includes bibliographical references (p.) and index.
ISBN 0-520-22366-7 (cloth : alk. paper).
1. Powell, David C., 1927– 2. Aquarists—United States—Biography. 3. Public aquariums—United States—Anecdotes. I. Title. II. Series.
QL31.P69 A3 2001
597.177'073—dc21 00-055169

Manufactured in the United States of America

10 09 08 07 06 05 04 03 02 01
10 9 8 7 6 5 4 3 2 1

The paper used in this publication meets the minimum requirements of ANSI / NISO Z39 0.48-1992 (R 1997) (Permanence of Paper).⊗

To my wife, Betty,
and our daughters, Eve and Amy,
and to all the animals
that made this possible

When once a man had known the sea and fished it, he was bound to it forever. When a man was no longer a part of it, through circumstance or even choice, there was always regret, and more often than not, envy of those who still belonged.

—D. R. SHERMAN
Brothers of the Sea

CONTENTS

Foreword, by Sylvia A. Earle *xi*
Acknowledgments *xv*

1 Underwater Thoughts *1*

2 Marineland of the Pacific *15*

3 A First Look at Realism *28*

4 The Road to Gonzaga Bay *42*

5 The Steinhart Aquarium *48*

6 San Francisco and the Sea of Cortez *54*

7 Sea World and South *66*

8 Carnival in Mazatlán *87*

9 The Lure of Sharks *103*

10 Tanner Bank and Mexico Expo *125*

11 The Revillagigedo Islands *131*

12 Roundabout to Steinhart Aquarium *146*

13 Search for a Living Fossil *163*

14 To Chile, Easter Island, and Rarotonga *175*

15 Monterey Bay Aquarium *184*

16 Creating the Exhibits *191*

17 Aquarists at Work *205*

18 Collecting the Fish *216*

19 Always Something New *238*

20 The Open Ocean *250*

21 Pelagic Fishes *269*

22 A Million-Gallon Fishbowl *284*

23 A New Direction *303*

Selected Reading *323*

Index *325*

FOREWORD

Sylvia A. Earle

In this long-awaited memoir, *A Fascination for Fish: Adventures of an Underwater Pioneer,* David Powell echoes the spirit of writer John Steinbeck and marine biologist Ed Ricketts, with a special brand of understated humor and matter-of-fact flair that friends will recognize as 100 percent Powell. There are answers here for millions of people who, like me, have stood in wonder in the Monterey Bay Aquarium before the magnificent captive forest of kelp or have caught their breath as they approached the display there of a slice of open ocean, shimmering with tuna, sharks, and bonito. "How is this possible?" I have heard people say. "How can big fish and live plants and small creatures, too, be brought here and seem so much at home?" Often I have wondered myself.

Here Powell provides glimpses of some of the everyday challenges in making these apparent miracles possible—coaxing great white sharks to eat while being careful not to be eaten; devising special pumps and other systems that aren't found in local hardware stores; creating artful lighting and other tricks to make walls disappear, giving the illusion that fish and other creatures confined in tanks are actually free in the sea. Most of all, he makes it clear that to make wild creatures prosper in captivity—whether octopuses, jellies, sand dollars, or sharks—

it helps to imagine yourself in the animal's place: literally, to "think like a fish."

Half a century ago, Powell began to work his magic at Marineland of the Pacific, attempting, as he puts it, to "give land-dwellers a glimpse into the underwater world." This man, who as a boy was so fascinated with fish that he slept with one under his pillow and who later became the highly respected wizard-behind-the-scenes at several of the world's finest aquariums, has done much more than create wondrous living exhibits. It may not have been his original intent, but he can rightfully claim credit for inspiring in several generations of terrestrial primates a sense of caring that comes with knowing what's out there, down there. Beyond his legacy of living artistry, technical breakthroughs, and pioneering exploration, Powell has had much to do with changing the way people regard the waters of the world. His ability to capture the essence of living coral reefs, sandy plains, rocky shores, and even the open sea and other wild places, complete with their amazingly diverse residents, has had a significant impact on the way people think about the ocean, about the beauty and wonder of life, and about themselves.

For many, the word "fish" brings to mind succulent morsels, lightly browned and swimming with lemon slices and butter; but those who see fish as presented by Powell—alive and well, swimming in beautiful places—are given another perspective, one that leads to an understanding that fish, like cats, dogs, horses, and birds, have individual quirks and even "personality." Moreover, when people see fish living in a near-to-natural setting, it begins to dawn on some that fish are to the sea as birds are to the land: vital components of the ecosystems that make the world function properly, the distillation of hundreds of millions of years of fine-tuning that now provide the underpinnings of earth's life support system.

When I first visited the Steinhart Aquarium in San Francisco in the 1960s, I found a number of people crowded around one tank, mesmerized by a gathering of small red fish perched among rocks encrusted with translucent red anemones, red soft corals, red sponges, and live red coralline algae. Nearby was a comfortable-looking abode for a spectacularly paisley-patterned but grumpy-looking scorpionfish, as well as other similar mini-marvels that would look in place in the Louvre or the Museum of Modern Art—or in the ocean realm from which they had been lifted for people such as me to see, up close and personal. This was my introduction to the handiwork of David Powell.

Experiences that have shaped the man and his work range widely in aquatic realms from distant reefs in the South Pacific to the deep sea offshore from Monterey. I have watched him in action often, including times we shared at the California Academy of Sciences and Steinhart Aquarium in the 1970s, but nothing compares with one occasion when I witnessed his fish-thinking wiles in action, with spectacular results.

During an expedition to the fabled "Islands of the Moon," the Comoros in the Indian Ocean, we chose to dive along the craggy volcanic shore at night in hopes of finding and capturing the notorious "flashlight fish," a small creature with bright patches of bioluminescent bacteria glowing under each eye. "No dive lights," said Dave, thinking like a fish. "Our lights would spook them. If they're there, we'll see the glow and we can just swim toward them." Following Dave into a submerged lava tube bristling with long-spined urchins, I watched as he carefully eased first one, then another small fish into his clear plastic collecting bag. Within a few minutes, Dave was transformed into an underwater Diogenes, with dozens of lanternlike fish illuminating our passage back to the tube's entrance.

In the pages that follow, Powell inspires laughter, wonder, spine-chilling concern, and philosophical musings as he recounts what it has taken to become the acknowledged grand master of wet animal husbandry. Although his narrative is based on a lifetime of experiences, starting in South Africa, where he was born in 1927, and spanning the major lands and seas of the world, his greatest strengths and best stories are drawn from decades of exploring North America's wilderness ocean off the West Coast, with special reference to the waters of Baja California and Monterey Bay. His accounts of diving in places along California's central coast when abalone were abundant and people were rare will astonish those who have arrived since.

I was particularly intrigued with Powell's description of a thoughtful discussion he had with underwater explorer Jacques Cousteau.

Though an expert spearfisherman personally responsible for the demise of countless hapless fish and other sea creatures, Cousteau awakened over the years to an awareness of fish that transcended their culinary appeal. Aquariums, he said, cannot do them justice; once I heard him refer to them as "fish prisons." Yet Powell believes, as do I, that the films Cousteau offered as an alternative, inspiring and wondrous though they are, cannot convey the thrill of an eye-to-eye encounter.

I witnessed this within my own family as my mother, an ardent fan of natural history films, sat in her wheelchair gazing into the great kelp tank at the Monterey Aquarium. Skeins of silver anchovies danced among golden-brown fronds like a flock of songbirds; a great starfish eased one arm into a new position; a large wolf eel slithered along the kelp's tangled holdfast; and then a large orange and black wrasse turned and seemed to look directly into my mother's eyes. "I had no idea it was this beautiful," she said. I am not sure whether she meant the fish, the kelp, or the scene as a whole, but I knew she could never step into the cold California waters just offshore and get to know the creatures there on their own terms. For her and for many millions of others who cannot or will not venture directly into the "big aquarium"—the sea itself—observing the microcosms that aquariums provide is the best way there is to experience and personally savor the nature of the living sea.

It has taken many people many years and substantial resources to make functioning realities of the institutions celebrated in this volume. David Powell pays generous tribute to all, neglecting only, perhaps, to put into adequate perspective his own substantial contributions. It is my pleasure, then, to encourage readers to look between the action-packed lines and see the enduring legacy, the spirit, the vision, and the ethic of this remarkable man who seems to have been having a wonderful time while making a living, making history, and making a difference doing what he loves to do.

ACKNOWLEDGMENTS

Whatever I may have achieved over the years has been due in part to the help of countless people who supported me and believed in what I was trying to do. Thanks, first of all, to my fellow aquarists and collectors, who participated in many an adventure and share my fascination with the world of water and its wonderful inhabitants. Thanks also to those too-often-unappreciated round-the-clock staff who keep the life support systems running and our delicate charges alive. To my understanding and supportive bosses at various institutions, Ken Norris, George Millay, John McCosker, and Julie Packard, thanks for giving me the freedom to play with ideas. Thanks, too, to Jim Hekkers and Nora Deans for convincing me that this book was a good idea—though there were moments when I wondered—and most of all to Sally Hekkers, whose unflagging encouragement kept me going and who patiently helped make sense of the disorganized words I was putting down. There would be no book without her.

Finally, special thanks to my wife, Betty, for letting me follow my dreams, no matter how crazy they may have seemed at the time.

1

UNDERWATER THOUGHTS

THE MORNING FOG WAS BEGINNING to lift as Bob Kiwala, Mike Weekley, and I headed out of Monterey Harbor in the eighteen-foot outboard. We were in our wet suits, and the boat was loaded with six scuba tanks and our dive gear as well as a gasoline-powered pump and a hundred feet of garden hose. The Monterey Bay Aquarium was nearing completion, the exhibits were coming together, and we had started collecting the thousands of fishes and invertebrates the visitors would be coming to see in the fall of 1984, now only a few months away.

As unlikely as a hundred feet of garden hose may seem, it was the key to success in collecting the beautiful burrowing sea anemones (*Pachycerianthus fimbriatus*) that are so abundant in the sandy mud bottom near the base of Monterey's Coast Guard breakwater. After dropping anchor in fifty feet of water, Bob and I put on our weights, tanks, fins, and masks, rolled over into the water, and headed for the bottom clutching one end of the hose. On board, Mike fired up the pump engine, and water jetted out of the hose end. Swimming down, we were buzzed by a couple of curious sea lions—who were wondering, no doubt, what we were up to.

On the bottom we saw dozens of sea anemones, their crowns of tentacles gently waving in the slight current a foot or so above the seafloor.

Their delicate look was deceptive, however, for not visible were the two-foot-long tubes that lie buried beneath the firm mud bottom. That's why we needed the pump and garden hose.

Although all of these anemones are the same species, they come in three distinct colors. Most have light gray tentacles, but for some reason a few have either deep purple or glowing orange tentacles. These orange anemones are puzzling. Because seawater, as it increases in depth, selectively filters out the red end of the light spectrum before the blue, no orange light reaches fifty feet below the surface; yet there they are, glowing bright orange. We understand the physics of how they appear orange when there's no orange light—the tissues in their tentacles refract the short blue wavelength and change it to the longer wavelength of orange—but the biological reason for this phenomenon remains a mystery. In any event, the visual effect is striking. Selecting a fine-looking orange specimen, I started blowing the muddy bottom away from around the anemone's tube with the jet of water from the end of the garden hose. Gradually working my hand and the hose deeper and deeper until reaching the bottom of the animal's tube, I gently lifted it out, unharmed, and placed it in a plastic bag.

What a bizarre way to make a living! Here I was, fifty feet below the surface of the bay, totally obscured in a cloud of mud, with my left arm stuck clear up to my armpit in the bottom, trying to dig out a sea anemone. In spite of the numbing cold and zero visibility, I was thoroughly enjoying every minute and felt quite at home in this underwater world.

At times like these, I wonder how I was ever so fortunate as to end up in this strange line of work. Yet looking back, I can spot clear signposts steering me away from well-traveled, traditional, acceptable employment routes straight toward my chosen career.

SLEEPING WITH FISH

The earliest sign, which I can't even remember, is one my mother told me about. I was born in South Africa in 1927 to American parents. When I was about five, my mother and father took my sister and me to the seashore town of Durban for a day at the beach and to do some fishing.

The author at age five, with sister Marcia, catching his first fish at Durban, South Africa. (Photo by Harold Clark Powell)

With my father's help, I caught a fish, perhaps my very first. That night I insisted on sleeping with it under my pillow, and my tolerant mother agreed. While this may have been on the fringes of accepted behavior for "normal" kids, I see it as an omen, the sign of a born biologist. After all, Jane Goodall, or so I heard, slept with earthworms under her pillow.

My family moved to England in 1932. As a boy growing up in England, I was constantly drawn to water and would catch newts, water boatmen, and dragonfly larvae to keep in jars in my room. Using a primitive microscope, I would study with fascination the beating heart and developing brood of babies inside the tiny female water flea (*Daphnia*). Off I would go on my bicycle with my fishing rod, dip net, and jar to explore the streams and lakes within a wide range of home or school. This was the early 1940s, and England and Germany were at war. Practically all food except bread and vegetables was strictly rationed, and my fishing efforts, while fun for me, contributed significantly to our table.

My first job, at the age of fourteen, was at a canoe and punt rental operation next to Oxford University's Magdelan College on the river

Thames. I kept the boats clean, bailed out water, and chauffeured people on leisurely cruises in a punt, one of those long, narrow boats propelled by pushing against the river bottom with a long pole. Many times during those years I read and reread Izaak Walton's *The Compleat Angler*, written in 1676, and relished both the fish lore and the poetry.

My mother had a tendency to embrace whatever was the latest educational theory. As a result of her avant-garde thinking, my sister and I attended a succession of drastically different schools: a nudist school where, weather permitting, teachers and students were naked; a school in Wimbledon with a philosophy similar to that of the well-known Summerhill, the only rule being not to hurt anyone; and a strict Church of England school.

Fortunately, I ended up in a fine Quaker school, though by then I had become something of a rebel and troublemaker. The Friends' School in Saffron Walden was in the southeast of England. At that time, the Battle of Britain was at its height, and planes from Germany, England, and America were being shot down, some within easy bike-riding distance of my school. Before long I began experimenting with explosives taken from the machine-gun cartridges of downed planes. One night an accomplice and I set off a homemade "bomb" in the school's goldfish pond. A geyser of water and goldfish shot twenty feet into the air. Frantically scooping up a few of the fattest ones, we ran off before the school staff discovered the origin of the loud noise. Sneaking into the chem lab, we cooked the goldfish in a pan over a bunsen burner. To perpetually hungry, growing boys enduring wartime food rationing, these fish, seasoned by the excitement of the adventure, tasted delicious.

My experiments with explosives grew in magnitude until a concerned classmate tipped off the headmaster. I was called to his office, given a stern lecture on the dangers of my activities, and expelled from school for a month.

Another sign of my biological bent came when, as a sixteen-year-old, I began to question some of the school's curriculum. What earthly use was French or calculus going to be to me when I was out in the world earning a living? I wanted to quit school and get a job somewhere working with fish. The understanding headmaster talked me out of it. Acknowledging my desire, he explained that without some sort of degree

I would end up cleaning fish tanks for the rest of my life. Years later, I had to laugh. Having earned a master's degree from the University of California at Los Angeles (UCLA), I found myself doing just that: cleaning fish tanks along with my other tasks at Marineland of the Pacific.

At age seventeen I graduated from the Friends' School and was now subject to the military draft. Fortunately, I had both American and British citizenship, so I signed up as a mess boy on an American freighter returning to the States in convoy, under nightly attack by German U-boats from Liverpool to New York. I spent the remainder of World War II on Liberty ships in the Pacific ferrying war supplies to the battle zones of Guadalcanal, New Guinea, and the Philippines. When the war ended, I served as a medical technician in the U.S. Army for a year and a half on Okinawa.

This military service qualified me to take advantage of the G.I. Bill's free education for service personnel. I enrolled at UCLA, finally able to focus on my dream. Although the U.S. government covered school fees and books, I had to pay my own living expenses. I therefore took on a wide variety of part-time jobs: I worked the night desk at a Sunset Boulevard motel and the graveyard shift at a gas station; I served as night attendant in the psychiatric ward of the Veterans Hospital—a very disturbing job—and maintained the live Maine lobster holding facility at a Malibu restaurant. For two summers I worked as a seasonal aide for the California Department of Fish and Game catching, measuring, tagging, and releasing yellowtail from sportfishing boats out of Long Beach and San Pedro harbors. It was great work, but it also gave me a taste of the bureaucracy in a government agency.

DESCENT INTO A NEW WORLD

I can blame Jacques Cousteau for nudging me in the direction that was to become my life. A friend gave me a copy of Cousteau's first book, *The Silent World.* I was so fascinated with his descriptions of his experiences underwater that I knew I had to learn to dive. Because I was a student and constantly short of money, though, diving was not a simple matter. I talked my younger cousin, Norman Powell, into going halves with me on whatever equipment we needed to get started.

PRIMITIVE DIVING GEAR

By present standards the diving equipment in the 1950s was primitive, hard to use, and potentially dangerous if you didn't keep your wits about you at all times.

The Cousteau-Gagnan regulator had two large hoses for the air, one for inhaling and one for exhaling. Although the regulator was quite easy to breathe with underwater, a major problem occurred any time you took the mouthpiece out of your mouth while you were in the water. The hoses immediately filled with water, including the one that was supposed to supply you with air. The air hose now became a water hose. If you stuck the mouthpiece back in your mouth and sucked, you got half a gallon of seawater instead of air.

The trick to avoiding that unpleasant situation was a little routine that had to be followed religiously. Whenever your air hoses became full of water—which happened rather often—you rolled over on your left side and exhaled. If you did it correctly, this blew most of the water out of the exhaust hose. The water that was in the air supply hose, on the right, now flowed downhill into the theoretically empty exhale hose. Inhaling very cautiously, you could get a lungful of air—mixed, of course, with a little of the water that didn't quite get blown out. The whole procedure was annoying at best, but downright dangerous if your mouthpiece was knocked out underwater.

Luckily, the double-hoses-full-of-water situation didn't last very long. An independent and inventive diver devised a pair of check valves that you could buy and install to keep the water out of the inhale hose. Now if it flooded, the only water would be in the small space in the mouthpiece itself. This made diving more pleasant and took much of the worry out of losing your mouthpiece in awkward places, like a hundred feet down or in the dark of night with your hands occupied with a lobster, dive light, and bag. Eventually, the U.S. Divers Company incorporated check valves in all of their regulators.

Primitive 1955 dive gear: a leaky "dry" suit, a homemade weight belt, and a Cousteau-Gagnan double-hose regulator. (Photo courtesy Betty Powell)

My first few dives were made in the summer, and I wore long underwear for "warmth." Maybe it helped, but not much. Thirty minutes into the dive I was shivering: I needed some way to keep warm. There were dry dive suits available, and I saved up for one made by Pirelli, the Italian car tire manufacturer. It was made out of thin, smooth rubber to keep the water away from your body, and you wore clothing underneath it for warmth.

The name "dry suit" was a misnomer because on practically every dive, whenever you brushed against a rock, a hole was poked in the thin rubber. The suit took on water, and there went the insulation from the cold. Climbing out of the water after a dive, I would feel the water that had leaked in run down to the lowest point, and my leg would balloon out as though I were suffering from an advanced case of elephantiasis.

After a couple of years, closed-cell foam neoprene came on the market and we had access to true wet suits. I couldn't afford a custom-made suit, so my wife, Betty, and her cousin Frank Parker's wife, B.J., measured me and cut out the rubber material to fit, and we all glued it together. I wouldn't call it the best-glued suit in the world, but it was a huge improvement over the perpetually leaking "dry suits."

We went to the only dive shop in California at the time, the French-run U.S. Divers Company in Westwood Village. We bought one tank, one regulator, a pair of fins, a Squale mask, and a little instruction pamphlet that came with the French-manufactured Cousteau-Gagnan Aqua Lung regulator, serial number 106. Not having access to a pool to try it out, the two of us putted out in my little ten-foot skiff to the Long Beach breakwater in search of some reasonably clear water. Nervously, I put on the tank, fins, and mask and slipped into the water. It was incredible! I was actually breathing underwater, and all around me and all over the rocky bottom were wonderful undersea creatures. I was truly in their world.

The Los Angeles Harbor is far from the most scenic dive spot in the world, but to me it was thrilling. After using up what I guessed was half the tank of air, I surfaced and clambered out on the breakwater rocks. Norman went next but was not nearly as impressed. He never did take to the underwater world like I did; eventually, in fact, he became a land-based geophysicist rather than a marine biologist. To each his own.

For the next year or so I went diving every chance I had. Most of my diving was done solo because I knew few other divers then. One year I was fortunate enough to get a part-time job collecting specimens for the invertebrate zoology classes at UCLA. This gave me a chance to dive and actually call it work, although the pay was very low. With my background in biology, I was fascinated with the life I saw beneath the surface, but I was equally excited by the very act of diving itself. However, just being able to breathe underwater in this new world and to stay submerged for relatively long periods of time didn't quench my thirst. I was eager to learn the secrets of the lives of each creature I saw—and for many years almost every dive I made revealed something new to me.

Although I was only one of the millions of people living in the vast, asphalt-covered city of Los Angeles, I knew I was one of the privileged few. I was able to enter an ocean full of strange, alien beings that exists right at our doorstep. Meanwhile, those millions went about their daily lives totally unaware of this fascinating world just beyond the shore.

One day one of the French divers at the U.S. Divers store mentioned that he had gone out the night before with an underwater light and caught some spiny lobsters (*Panulirus interruptus*). That sounded pretty exciting; besides which, it had great gastronomic possibilities. As a working college student I couldn't afford to buy luxuries like lobsters, but this could be a way I could get some virtually free, just for the picking. The diver showed me the light he used. It was a basic two-cell flashlight inside a specially made rubber case, with a hose clamp where the rubber fits around the plastic lens. Of course, the dive shop wanted more money for the light than I was willing or even able to pay.

At the time, I was working the graveyard shift in a gas station, and we quite often replaced burned-out headlights on cars. A car headlight has two filaments, for the low and high beams. When the low beam burns out, the lamp has to be replaced, even though it still has one perfectly good beam. I suspected those lamps had just the kind of brightness I needed for my first underwater venture. What's more, the sealed headlights were designed to resist both water and pressure. All I needed was a big enough battery and a long, heavy electrical cord, and I'd have myself a remarkably serious underwater light: a car high beam underwater. At a war-surplus store I found fifty feet of heavy-gauge, well-insulated, very cheap electrical cord. After soldering the two wires securely to the lamp terminals, I waterproofed the connection with some tar-impregnated electrical tape and then melted it together with a flame over the kitchen stove. When connected to the battery in my car, the light was really bright, and it worked quite well underwater in a bucket. I was in business.

I talked Norman into coming with me, and one dark evening we rowed my little plywood skiff from the Santa Monica Pier out to the harbor breakwater. Decked out in tank, mask, and fins, I rolled over the side into the black water. Taking the light from Norman, I aimed it downward—and saw nothing but murky water, the bottom nowhere in view since it was beyond the beam of the light. It was spooky to be surrounded by nothingness, and for a moment I wondered if I was totally nuts to be doing this. I swam down cautiously through the dark, and finally rocks came into view, and on them was all the marine life I was familiar with. Now that I was in their reassuring presence I felt comfortable and began to look around.

There were two-spot octopuses (*Octopus bimaculatus*) cruising the bottom for food, and fishes huddled for the night in crevices between the boulders. A sheep crab (*Loxorhynchus grandis*) drew up its claws defensively when the bright light hit it. Soon I spotted the first lobster crawling over the boulders: its red color stood out beautifully in the bright car headlight. Apparently my presence or the light made it nervous, though, and when I made a grab for it the lobster shot off backward with amazing speed and disappeared into the dark. Whoa! This was not going to be as easy as I thought! Lobsters don't just sit there waiting to be picked up. Cruising on a little farther I spotted another one, and this time I grabbed it with all the speed I could muster. It worked; I now had a firm grip across the back of the frantically flapping lobster.

Returning to the surface I found the skiff and Norman, who had the challenging job of trying to follow my bubbles and the electrical cable in the dark. I tossed the lobster over the gunwale into the boat. Back down I went, and by the time I ran out of air we had six nice lobsters crawling around the bottom of the skiff, the largest weighing about eight pounds. This night diving was all right! For seventy-five cents' worth of scuba air and our time, which in those days was dirt cheap, we had a gourmet feast for a number of people.

BUILDING A TICKET TO EXPLORE

The little ten-foot skiff with its often unreliable seven-horsepower outboard engine had proved to be fine for diving in calm, nearby areas with a maximum load of two pretty lightweight people, but I was anxious to see what lay farther afield. The coasts off the Palos Verdes Peninsula and Catalina Island were well outside the safe range of my little skiff. Norman and I decided to build a boat that could take us to those exciting places twenty miles away.

Norman was working at the time for the Los Angeles Department of Water and Power and had a reliable income. I was still going to UCLA plus working part-time, and I had little money to spend on anything beyond bare necessities and the cheapest California jug wine. Norman generously offered to buy most of the equipment and materials; I would contribute my labor. We decided on an eighteen-foot plywood hull that would be powered by a seventy-five-horsepower, four-cylinder inboard marine engine. Buying materials and equipment a little at a time, we worked on it in our spare time for well over a year.

For me the experience meant an ongoing series of problems to solve, not the least of which was developing the self-discipline to keep focused on the dream of the completed boat. There were times when I became discouraged by the seemingly endless tasks that lay between the present and our goal. The hundreds of problems to solve during the construction had their own small rewards, though, and with the completion of each one there was a sense of accomplishment.

As rewarding as the creative process and the end result were, it's not an experience I'd want to repeat. Building a boat is about ten times more work than you think it could possibly be. By this time I had owned three other small boats, and I knew I would always have a boat in my life. Part of me must be a soul mate of Ratty in *The Wind in the Willows,* who said, "Believe me, my young friend, there is *nothing*—absolutely nothing—half so much worth doing as simply messing about in boats." In later years, though, when I was no longer broke, I would buy used ones and fix them up rather than build from scratch.

The day finally came when our boat was done and we launched it from the Santa Monica Pier. It was a pleasant surprise when everything worked as it was supposed to. Giving throttle, we planed along at a respectable speed of twenty knots. What a joy it was to finally experience the culmination of our months of work! I ended up using the boat far more than Norman did. When I came into a little inheritance money after my father's death a few years later, I took it all and bought out Norman's share.

Actually, I give that little boat a great deal of credit for my career in public aquariums. It provided me with the opportunity to dive in a host of interesting places and to learn firsthand about the life in the underwater world. The observations I was able to make of the habitats and behavior of marine animals, not to mention the diving skills

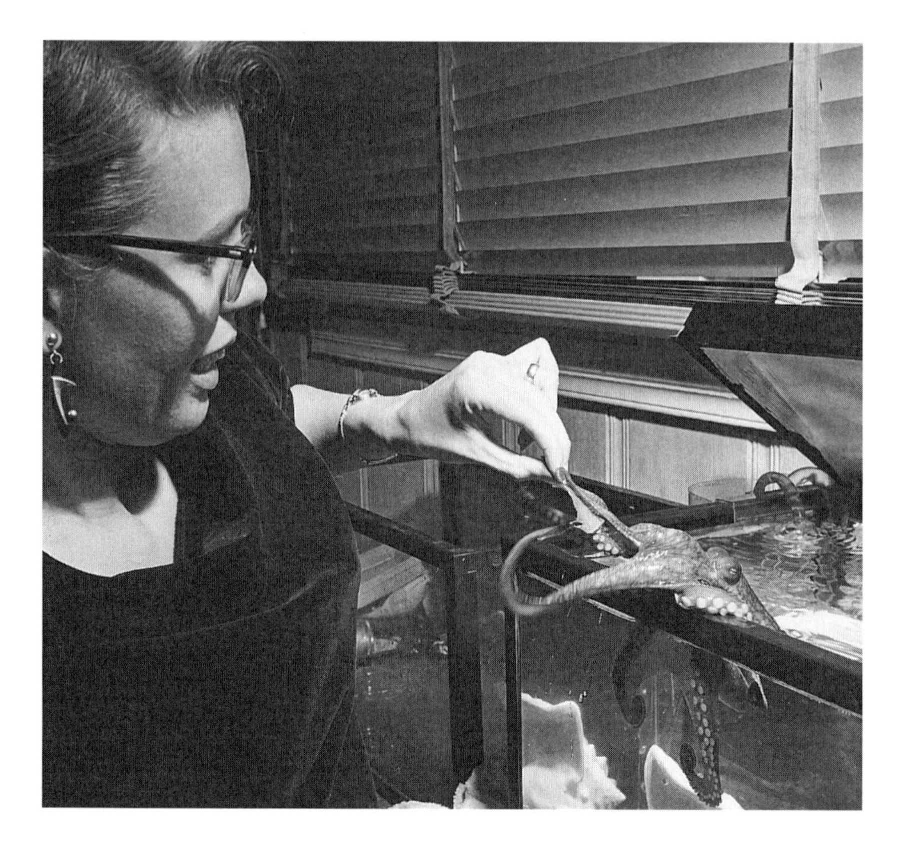

Betty makes friends with my octopus. (Photo courtesy Don English)

I acquired, were invaluable to me. In addition, that boat, plus the circumstances of being constantly poor, gave me a do-it-yourself attitude and taught me the art and skill of problem solving, improvising, and inventing. These hard-won lessons would serve me well for years to come.

OCTOPUS IN MY LIVING ROOM

Early on I set up a small marine aquarium in my room to keep some of the creatures I collected while diving or tide pooling. One creature in particular, a little octopus, fascinated my friends. One day in 1954, Betty Mumby, a fellow UCLA student, asked if she could come up and see my octopus. That was probably one of the more unusual lines to lead to a relationship, but it certainly worked. We started seeing more and more of each other, and pretty soon we married.

Betty worked full-time in the UCLA Admissions Office while I finished my master's degree. A couple of years after we married I finally got my degree and started to look seriously for a job. By then Betty was seven months pregnant. Unfortunately, the world wasn't crying out for marine biologists, and I couldn't find a thing. I even applied for work in the aircraft industry—twisting the truth about my education just a bit—in an effort to get any kind of paying job. Betty, now very close to term, finally quit her job on a Friday, I found a job on Monday, and our daughter Eve was born on Thursday. That was cutting it pretty close.

In desperation I'd walked in the front door of the Hyperion Sewage Treatment Plant and asked if they had any job openings. I could barely believe it when they told me they had a temporary job in the lab. That huge plant treated and discharged all the sewage from the City of Los Angeles. My job was to test for coliform bacteria on the seawater samples taken daily in Santa Monica Bay from Malibu to Palos Verdes. Once I got used to the odor, it was a pretty good job. We sewage workers had a saying: "It may be just shit to you, but to us it's our bread and butter."

The job had an interesting fringe benefit that was right up my alley: once a month they would collect fishes and invertebrates from the bay to determine if the sewage and the sludge discharging were having any effect on marine life. A series of stations at different depths from fifty to six hundred feet were sampled with a trawl net dragged across the bottom. I got to go out on the boat every month to help catch, count, and identify the creatures that came up in the net.

I was fascinated with some of the invertebrates that lived in the depths of the bay—grotesque, long-legged crabs covered with spines (*Paralithodes* spp.), delicate branching corals, and sometimes a whole netful of fragile pink sea urchins (*Allocentrotus fragilis*)—and I would take some of the animals from the shallower samples to keep in my home aquarium. The word quickly spread among my fellow members of the Marine Aquarium Society of Los Angeles, who were eager to see the strange and wonderful creatures in my aquarium. I can credit this hobby, my diving, and our little boat with landing me my first job as an aquarist. And what a job it was!

2

MARINELAND OF THE PACIFIC

A TIP IN 1958 FROM JERRY FAWCETT, president of the fledgling Marine Aquarium Society of Los Angeles, let me know that Marineland of the Pacific was looking for an aquarist. Marineland, which opened in 1955, was the world's second oceanarium, following on the success of the original Marineland in St. Augustine, Florida. Bearing the newly coined name "oceanarium," these two facilities were totally different in concept from the classic indoor aquarium. Located where year-round weather was mild, most of the main exhibits were outdoors. Marineland of the Pacific, perched as it was atop the cliffs of the Palos Verdes Peninsula in Los Angeles, had a spectacular view of the Pacific Ocean and Catalina Island on a clear day.

A phone call to curator Ken Norris led to my first job as a professional aquarist—my real break into the field of aquatic biology, on which I had set my heart as a boy in England. It was what I had prepared for, and I was ready.

I saw this as my opportunity to share with others the fascination I had for the creatures of the sea: to collect animals and create exhibits that would give land dwellers a glimpse into the underwater world. I didn't think of it as a profession—although it turned out to be one; I just knew it was what I wanted to do. My formal schooling was behind me, but the lessons I was to learn about marine life were just beginning and would continue for the rest of my career. I would learn about the fascinating behaviors of a host of animals and in so doing develop a profound respect for them. In time I would develop methods for collecting a whole range of animals safely and unharmed, knowing that it was my obligation to care for them as best as I could. A new world lay ahead of me.

Ken showed me around the place, pointing out the smaller aquariums, called "Jewel Tanks," that would be my responsibility. These exhibits were a rather haphazard mixture of freshwater and marine, local and exotic, fishes and invertebrates and seemed to be randomly scattered throughout the oceanarium complex. Marineland's star attractions were the performing marine mammals—the bottlenose dolphins (*Tursiops truncatus*), Pacific white-sided dolphins (*Lagenorhynchus obliquidens*), and California sea lions—but the facility also had a half-million-gallon display tank of fishes and sharks. Most of the Jewel Tanks weren't much larger than good-sized home aquariums, but a couple, such as the Octopus Grotto and the Hawaiian Reef, were considerably bigger, holding about ten thousand gallons.

Comparing the 1950s Marineland of the Pacific to the aquariums of the present day reveals striking changes. Those were the days before fiberglass tanks, acrylic windows, plastic pumps, and silicone sealant. The technology to make curved windows or tank walls simply didn't exist. The smaller Jewel Tanks at Marineland had rectangular welded-steel frames with slate sides and bottoms and flat, glass windows. Aquarists had to constantly scrape and paint the perpetually rusting tank frames and reseal the frequently leaking windows.

I didn't appreciate the management style of Ken Norris at the time, though I certainly do now. He gave me the responsibility, support, and freedom to create what I believed in. He said, in effect, "Here are all the tanks; go to it. I'm here if you need help." He left me alone unless I needed his advice. He gave all those who worked for him the feeling that he believed in us, and this motivated me and, I'm sure, many others to do the best we could. I learned a lot by following Ken's example, and, most important, I learned not to be afraid to try things that had never been done before. He had a wonderful and

PIONEERING DOLPHIN RESEARCH

With Ken Norris as curator, Marineland was an active research facility in the early 1960s. It was already known that dolphins emitted rapid bursts of sound in the form of clicks that were believed to enable them to navigate and detect objects underwater. This was similar in function to the echolocation used by bats at night. Unknown, though, was how sensitive this sonar is, and how much the dolphins rely on it compared to sight. So Ken and his assistant John Prescott set out to scientifically and methodically investigate the sonar capabilities of dolphins.

A young female bottlenose dolphin named "Kathy" was the test subject. She was trained to accept soft latex rubber cups (modeled after "falsies") being placed over her eyes to function as a blindfold. Without her use of sight, Kathy would have to rely solely on sound. The training took considerable patience and time, but through positive reinforcement, Ken and John gradually gained her trust. Eventually, wearing her two latex blindfolds, Kathy was able to swim around her tank just as well as if she were seeing with her eyes. The speed at which she found mackerel tossed into the tank was no different when she was blinded than when she could see.

The next step was to find out just how sensitive her sonar was. Smaller and smaller pieces of food were used, and she still had no trouble finding them. She could even tell one species of identically sized fish food from another just by the sonar image she received. A tiny BB pellet was dropped twenty feet away on the far side of her tank, and she was able to quickly locate it. The conclusion drawn from these elegant experiments was that bottlenose dolphins have a truly extraordinary underwater navigation system that is at least as acute as their sight.

irreverent sense of humor and seemed to be having fun in everything he did.

Ken died in August 1998, and among the outpouring of tributes paid to him by students and colleagues, Lawrence Ford said: "Ken taught with love, but [half] of loving somebody [is] getting out of their way. So he would set up the situation that gave the students a sense of trust enough so that they could open themselves up to test out being creative. . . . Then he got out of their way." He was an extraordinary human being who had a powerful influence on all he came into contact with.

PIONEERING BREAKTHROUGHS

At Marineland we made many breakthroughs in the collection, husbandry, and exhibition of marine animals. Under the creative direction of Ken Norris, together with biologists Bill McFarland, John Prescott, and Don Zumwalt, aquarist Jerry Goldsmith, and the highly skilled and knowledgeable collector Frank Brocato, we were able to present to visitors animals never before seen in aquariums. Nothing seemed too big a challenge. Marineland staff even captured the first whale ever to be displayed in an aquarium. A one-ton female pilot whale (*Globicephala macrorhynchus*) named "Bubbles" was trained to participate in the marine mammal performances alongside the bottlenose dolphins and the beautiful Pacific white-sided dolphins.

The main exhibit of fishes was housed in the half-million-gallon Oval Tank, a hundred feet long by fifty feet wide and open to the sky. Visitors could view from above and on two levels underwater through relatively small, double-pane, tempered-glass windows evenly spaced all around the tank.

Frank Brocato and "Boots" Calandrino did a remarkable job of collecting a wide variety of California game fishes for the Oval Tank and other exhibits. As former commercial fishermen, they had learned the habits and behavior of many kinds of fish. But there is a crucial difference between catching fishes for the market and bringing them back alive and healthy for an aquarium. Ken, Frank, and Boots therefore worked hard to develop new types of equipment that would catch fish without injury.

Spectacular synchronous jumps of Marineland's Pacific white-sided dolphins. (Photo Marineland PR)

The result, after much trial and more than a few errors, was an exhibit stocked with an impressive collection of fishes. The mix eventually included several giant black sea bass (*Stereolepis gigas*) weighing three to four hundred pounds and many large white sea bass (*Atractoscion nobilis*); leopard sharks (*Triakis semifasciata*) and bat rays (*Myliobatis californica*); fine schools of bonito (*Sarda chilensis*) and barracuda (*Sphyraena californica*); as well as countless smaller species like kelp bass (*Paralabrax clathratus*), sand bass (*P. nebulifer*), and sargo (*Anisotremus davidsonii*).

Marineland divers, working under the direction of former commercial diver Jake Jacobs, put on regular underwater feeding shows. The commercial helmet diving gear they used was heavy and seemed especially so as they walked on the bottom of the tank feeding the fishes. The gear was not easy to maneuver, and one had to be careful not to lean too far forward lest one topple over and lose the air-supplying helmet.

When Marineland first opened they had bottlenose dolphins in with the fishes in the Oval Tank. This proved to be a mistake. Being highly curious animals, the dolphins were constantly looking for stimulating activities. Unfortunately, this frequently involved the unwilling participation of the fishes that shared their tank. One activity demonstrated the ingenuity of the dolphins especially well. A few piles of boulders had been placed on the bottom of the tank as habitats for California moray eels (*Gymnothorax mordax*). When the ever-alert dolphins were cruising above, the eels knew better than to leave the security of their rock-pile home—something that frustrated the dolphins no end. Also in the tank were a few California scorpionfish (*Scorpaena guttata*), which rely on painfully venomous dorsal spines for protection. One unusually skilled dolphin figured out how to carefully catch and hold a scorpionfish in its mouth in such a position that the venomous spines pointed away from it. Armed with the hapless fish, the dolphin would then poke at the eels hiding in the rock pile until finally an eel panicked and dashed out of its shelter. At this precise moment the dolphin dropped the scorpionfish and grabbed the moray to play with like it was a brand-new toy. Experts say that humans, chimpanzees, Darwin's finches, and sea otters are the only tool-using animals, but this was a remarkable example of an inventive dolphin using a living tool for a new purpose.

One of Marineland's pioneering projects was the design and construction of a large fish transport tank that was carried on an eighteen-wheel diesel tractor-trailer rig. We used it for a major collection of fishes from Mexico's Sea of Cortez, including gulf groupers (*Mycteroperca jordani*), broomtail groupers (*M. xenarcha*), and the giant totuava (*Totoaba macdonaldi*).

The totuava, the largest of the croaker family (Scienidae), can reach a length of six feet and weigh over two hundred pounds. Abundant in the 1950s, the species is now severely depleted owing to overfishing and the reduction of its spawning grounds. So much water is being pumped from the Colorado River to supply the people of California and Arizona that little is left to supply the fish's breeding grounds at the mouth of the river in Mexico.

Always seeking even more exciting fishes for Marineland, the collecting crew made another expedition with the transport tank to San Blas, south of Mazatlán on the mainland of Mexico. The quarry was a pair of giant thirteen-foot sawfish (*Pristis perotteti*), which for many years made an impressive display in the Oval Tank.

These were exciting times, and all of us were caught up in the enthusiasm. It was also a learning period for people destined to move on to other places. Ken Norris completed his doctorate while at Marineland and left to become one of the founders of Sea World in San Diego, Sea Life Park in Hawaii, and Ocean Park in Hong Kong. Later he became an inspiring teacher and one of the world's leading marine mammal biologists at the University of California at Santa Cruz. John Prescott left to become the director of the New England Aquarium. Jerry Goldsmith hung on until Marineland's dying days before leaving to become the corporate curator of fishes for all four of Sea World's oceanariums.

EVOLUTION OF AN ORGANIZATION

It was interesting to watch the gradual evolution of the Marineland organization from its exciting early days to its demise in 1987. In the beginning, the staff wasn't afraid to take risks and try new projects. As the years passed, this attitude slowly changed and complacency set in. Management staff at Marineland seemed unable to visualize the returns from creating new exhibits. Being fascinated by the diversity of fish and invertebrate marine life, for example, I couldn't believe my ears when one day I heard the general manager say: "No fish is worth more than five bucks!" At the time, I put it down to his belief that the only reason visitors came was to see the performing marine mammal shows. Years later the Monterey Bay Aquarium was to show that one can achieve overwhelming success without having a single performing mammal show.

The gradual changes I saw at Marineland have happened at other institutions. Management becomes cautious and less willing to take the risks that come with innovation, focusing instead on variations of what's worked in the past. This passiveness in turn has a profound effect upon the institution's employees. It's sad to see the early enthusiasm and creativity wane, to be replaced by a "play-it-safe" approach.

In December 1965 Marineland missed a great opportunity to pioneer what was to become the most popular exhibit in oceanariums. Don Goldsberry and Ted Griffin, from Seattle, had netted and trapped a young, thirteen-foot female orca, or killer whale (*Orcinus orca*), in a narrow inlet near Puget Sound. Earlier they had captured a large adult male orca named "Namu." Griffin and Goldsberry used a floating pen to tow Namu to a larger pen they built off the end of a pier near downtown Seattle. Here Griffin had a remarkable one-on-one underwater experience with the intelligence and gentle nature of the big male orca. They befriended each other as Ted hand-fed chunks of salmon to the animal and they swam and dove together. He realized the killer whale's potential as an exciting oceanarium exhibit, and as a way to debunk the myth of the orca as the "vicious killer of the seas."

With the young female netted in a narrow inlet, Ted called Marineland and offered Bill Monahan, the general manager, the first opportunity to buy the young female whale. Monahan said, in effect, "No thanks, we have pilot whales bigger than that, so why would Marineland want to pay a lot of money for another whale just because it's black and white?" Ted hung up the phone and called George Millay, president of the new Sea World, down the coast in San Diego, and made the same offer. Millay snapped it up.

The rest is history. The whale, named "Shamu," was successfully airlifted to San Diego and soon became the star of Sea World's marine mammal shows. Visitors flocked to San Diego to see this beautiful animal perform. Needless to say, the boost in admission revenue quickly recouped Shamu's buying price. Marineland eventually acquired orcas, but they missed being the first.

INVENTION OF THE SLURP GUN

I was learning fast during the years 1960–61, but not only at Marineland. I had my boat tied up in a slip at San Pedro in Los Angeles Harbor, and I took every opportunity I could to use it. Owning a boat and being poor are not a good combination. I looked for ways to make a little extra money to help cover the expenses of the proverbial wood-lined hole in the ocean into which one pours money. Mine was a small hole, but still a hungry one.

I had continued to collect animals and keep them in aquariums and had established a small-scale business selling fish to pet stores. Most of the demand was for colorful tropical marines, but there was a potential market for anything with bright colors. I'd made the acquaintance of Dr. J. Walter Wilson, a Beverly Hills dermatologist and regular member of the L.A. Marine Aquarium Society. He had a special fascination for a little crimson native, the bluebanded Catalina goby (*Lythrypnus dalli*).

Although not a diver himself, Walter had an inventive mind and had designed a device he called a "slurp gun" specifically to collect this goby. The blue-banded goby is probably the most abundant fish around Catalina Island and can be seen everywhere perched on rocky reefs in twenty to eighty feet of water. They are always within darting distance of protective shelter, a behavior that makes them almost impossible to catch with a net. Their haven may be a tiny rock crevice, or if all the holes are occupied by other gobies, they hang out near the protective spines of the long-spined sea urchin (*Centrostephanus coronatum*).

Through trial and error and much testing, Walter and I perfected his slurp gun to the point where I could collect fifty or sixty gobies on one tank of air. The apparatus was made of a clear acrylic tube and powered by stretching and cocking a speargun rubber. When the trigger was pulled, the taut rubber and piston flew back, sucking the surprised fish into the barrel. We were both delighted that it worked so well.

In addition to supplying Walter with all the gobies he wanted, such trips allowed me to supplement my income by selling Catalina gobies to the few aquarium shops that sold marine fish. At fifty cents per goby, I could make a hundred dollars on a day trip to Catalina Island in my boat. This far from paid for the cost of keeping a boat, but it certainly helped. Unfortunately, the demand for the beautiful bluebanded gobies was limited; it didn't take long for me to flood the small L.A. market, and my goby income took a dive.

Walter, who thoroughly enjoyed designing and making gadgets, next built the Big Bertha of all slurpers. Six feet long and six inches in diameter, the acrylic plastic device was powered by four speargun rubbers. It was cumbersome underwater, but I was able to collect the ten-inch-long mantis shrimp (*Hemisquilla ensigera*) by literally sucking them out of their sandy seafloor burrows. When I spotted a shrimp's stalked eyes peeking out of a hole, I placed the clear end of the gun barrel over its burrow. I waited until the shrimp came up for another look, then quickly pulled the trigger. The powerful gun sucked a crater in the bottom and instantly filled with sand, shells, and a very irate mantis shrimp frantically dashing and whacking about inside the slurp gun.

Despite their common name, these strange creatures are not true shrimp. The mantis part of their name comes from their praying mantis–like front appendages. These are tools well adapted for seizing and dismantling prey, and they afford good protection from predators. The most remarkable feature of these appendages is their lightning speed: they shoot out faster than the human eye can follow. The lower part of their hinged claw is blunt and is used like a hammer for cracking hard-shelled clams. The extensible tip is like a needle and can pierce their prey, while the razor-sharp inner edge of the hinged arm can slice their prey in half. It's the animal equivalent of a Swiss army knife! Commercial fishermen know the mantis shrimp by the very apt name of "thumb splitter."

The effectiveness of mantis shrimp tools became vividly clear to me when, back on the boat, I scooped one up in my dive mask to transfer it to another container. It shot out its weapon and pierced clear through the tough rubber of the mask; unable to pull out, it just helplessly hung there. Another time, I was temporarily moving a mantis shrimp out of my home aquarium to clean the tank. Knowing something of their capacity to shred nets and fingers, I carefully scooped it into a large glass tumbler. Suddenly there was a loud crack and the glass shattered.

This species of mantis shrimp has two brilliant red, blue, and yellow false eyespots on its tail. Many widely differing animals, from insects to fishes, use false eyespots to confuse or startle potential predators.

A PAINFUL LESSON

I learned a great deal about the identification and relationships of fishes during Dr. Boyd Walker's intensive UCLA course in ichthyology. But there's something seriously lacking when all you see is the grotesque dead body of a pickled study specimen. Swimming with fish in their own world, collecting them, and keeping them in an aquarium environment are the ways to really understand these creatures.

For example, the California scorpionfish relies on potent venom in its spines for protection. Sedentary in their habits, these slow swimmers—when they do get up and move—are extremely easy to catch. Because of the ease with which they can be caught, there is a tendency to take them for granted. That can be a painful mistake.

In 1962 John Prescott and I were on a diving trip on board Marineland's collecting boat *Geronimo* to Santa Barbara Island offshore. We were diving in a kelp bed on the lee side of the island to collect kelp forest fishes and invertebrates. I hand-netted a nice large scorpionfish, and partly because of the strong surge I barely pricked my finger on one of its spines as I transferred it to John's net bag. It hurt, but not enough to make me want to quit the dive. When we were almost out of air we started swimming back to the boat against the current.

Swimming hard, John accidentally kicked the net bag that was trailing by his feet. His vigorous kick drove the scorpionfish's dorsal spines through his wet suit deep into the instep of his foot. The pain was instant and severe, but somehow he managed to make it back to the boat. Now the dive trip was definitely over. John's pain was intense, and he suffered agonies on the long trip back to San Pedro.

At the time we didn't know the remedy for scorpionfish venom. Through ignorance, we treated John's wounds with ice—exactly the wrong thing to do, we later found out, and it no doubt made John suffer more. We eventually learned that the most effective treatment is to immerse the wound in very hot water. Heat denatures the toxin, and the worst of the pain disappears almost immediately. John's ordeal was a dramatic demonstration of the effectiveness of intense pain as a defense. Few predators will mess with a scorpionfish more than once.

In the 1960s, sportfishermen and fisheries biologists became interested in establishing artificial reefs in flat, featureless sandy or muddy bottom areas. This was perceived as a way to increase available reef habitat that would then attract resident fish for the benefit of fishermen and divers.

In line with this philosophy, a number of old Los Angeles streetcars were dumped over a sandy area near the south end of Santa Monica Bay. A few months after they were installed I had a chance to dive on them with Earl Ebert and Chuck Turner, biologists with the California Department of Fish and Game. The underwater reefs had already attracted a healthy population of fish, and it was quite humorous to watch them swimming in and out through the windows or perched on the streetcar seats.

The streetcar reef was one of those projects that initially seemed like a good idea but in reality was not. Having seen the effect of marine animals on the untreated parts of wooden pier pilings, I knew these wooden streetcars would soon be devoured by the prolific underwater teredos, or shipworms—which, despite the name, are not worms at all but are wood-eating clams (*Teredo* and *Bankia* spp.) with long siphons—and by crustaceans known as gribbles (*Limnoria* spp.). All that would remain would be the rusting metal and the glass windows.

The streetcar reef did offer us at least one thrill. Over a nearby area of sandy bottom about a city block in size we came across hundreds of giant sheep crabs. In some places they were literally piled on top of one another. Both sexes were present, with the larger males having a leg spread of over two and a half feet.

It was the first and only time I have seen such an aggregation. Where did they come from, and why did they come to this particular place? Sheep crabs move rather slowly, and if they came from far away, which I assume they did, it must have taken them quite some time to get to where they were now. How did they know where they were going, and what was it that prompted this massive migration? Perhaps it was for reproduction. I had occasionally seen mating sheep crabs at other places, but it was always just a single pair, not a huge gathering like this. It still puzzles me.

Sheep crabs are a type of spider crab. When small, they have the interesting habit of decorating themselves, and for that reason they're also known as decorator crabs. Using their claws, they hook pieces of their surrounding environment to their bodies. The result is excellent camouflage that must protect them from being seen and eaten by large fishes. Decorator crabs use whatever happens to be around, such as seaweed or fluffy bryozoans or occasionally a few strawberry sea anemones. Unless the little crabs move, it is almost impossible to distinguish them from the surrounding growth. As they grow and become too large for most fish to swallow, they give up their decorating behavior and depend on their size for protection.

3

A FIRST LOOK AT REALISM

At every opportunity I was out collecting animals for either Marineland of the Pacific or my own home aquarium. The fish and invertebrate life I saw while underwater or in the rich tide pools of Palos Verdes Peninsula was much more impressive than what was represented in the comparatively bare and static aquariums at the oceanarium. For instance, Marineland had an exhibit of tide pool animals and plants, but unlike real tide pools with their dynamic waves and surge, it was motionless. To make things worse, the exhibit was in a square tank, a shape almost never seen in nature.

The intertidal pools at Portuguese Bend in Palos Verdes were formed by long fingers of rock jutting out from the shore into the ocean. Narrow channels between the rocks were home to a great variety of tide pool life, including fishes: young opaleye (*Girella nigricans*) and two types of tide pool sculpins (*Clinocottus* spp. and *Oligocottus* spp.). These animals have evolved to be quite at home in the waves that surge through the channels.

To us, the powerful waves that pound the intertidal rocks look as if they would sweep away everything in their path, but beneath the surface the fish take shelter in quiet eddies until the tide goes out and the period of calm water returns. The little sculpins have gripping pelvic fins that grasp the rough bottom. Some, like the clingfish (*Gobiesox meandricus*), have pelvic fins that have developed into a fully formed suction cup to hold them firmly in place amid the crashing waves.

Other fascinating creatures could be seen here if you looked hard enough. In the gravelly sand beneath certain rocks you might find the burrow of a pair of ghost shrimp (*Callianassa affinis*) together with a pair of their commensal burrow mates, the blind gobies (*Typhlogobius californiensis*). Blind gobies reap the benefit of a ready-made home built by the industrious shrimp, but what benefit, if any, the shrimp gain from the arrangement is not clear. Perhaps the little fish protect the shrimp from possible predators like the intertidal octopus (*Octopus bimaculatus*) by nipping at the tip of an exploratory tentacle entering their burrow.

My experience with the real intertidal world made me dissatisfied with the unrealistic representation of tide pool life we had in the Marineland exhibit. So I began thinking of ways I could make it more realistic, more like the tidal channels of Portuguese Bend with the succession of waves washing over them. Obviously I couldn't create a long, narrow tank in a space only four feet long, but that image lingered in my mind. Then, for some reason, the fun house on the Pacific Ocean Park amusement pier, and the illusions created there with mirrors, popped into my mind.

Why not create my tide pool channel with mirrors? I decided to give it a try and began sketching the positions and angles of the rocks and mirrors necessary for this idea to work. It was important that visitors not be able to see their own reflections looking back at them from either of the two mirrors. That would definitely ruin the illusion.

Fellow aquarist Jerry Goldsmith and I got to work. First we chiseled out the old rocks from the tide pool tank, and then, with the help of Jerry's creative eye and muscle, we collected new rocks of appropriate sizes and shapes. Eventually the backbreaking work of positioning the two mirrors and cementing the rocks in place was done, and the tank was filled. The concept worked, and what we formerly saw as the rear of the tank now looked like a long, narrow, rocky channel going back eight or ten feet. After the cement was cured, the fishes, invertebrates, and algae were added. The fish were a little confused by the mirrors but quickly figured out that what they saw reflected in them was not the real world. After a while, they ignored the mirrors and their own reflections.

Although the exhibit was a vast improvement, it still didn't look like reality, for a simple reason: it lacked motion. What was missing was a wave periodically crashing into it, just like real surf does in the pools at the shore. Manually dumping a bucket of water into the tank produced the right effect, but I couldn't keep doing that all day long. I needed an automatic bucket dumper.

Stealing a stainless steel bucket from the mammal trainers, I had a welder attach a pivot point to each side of the bucket just below its centerline. When the bucket was empty it hung there upright in normal bucket fashion, but when it was filled with water it became top-heavy and flipped over, dumping the water. Once empty, it righted itself and was ready to fill again. All we needed now was to suspend the bucket securely over the tide pool exhibit near the front window and supply it with a continuous flow of seawater.

Once installed, it looked terrific. The bucket tipped, the wave crashed down, and the tank turned white with foamy bubbles. The bucket righted itself and began filling again. The bubbles cleared quickly, and then you could see fishes swimming near the end of the long rocky channel. What people didn't realize was that the fish they saw in the back of the tank were reflections of the same ones they saw in the front. Only occasionally, when a sea star crawled up a mirror, was the trickery revealed.

The combination of the periodic wave action, the rockwork, and the two mirrors made a most effective tide pool exhibit.

FARNSWORTH BANK EXPERIENCE

By the late 1950s the popular (if rather corny) television show *Sea Hunt,* starring Lloyd Bridges, made the sight of a scuba diver less of an oddity than it was in 1952; scuba diving began to catch on, and a number of dive shops started up along the coast of California. These were not the first, though. Back in 1953, twin brothers Bob and Bill Meistrell had launched a dive shop called Dive n' Surf in the Santa Monica Bay town of Redondo Beach. They made the very first quality wet suits, and all of us Marineland divers had our suits custom made by them.

Years later their wet suit business evolved into the highly successful line of Body Glove products.

In the early years Dive n' Surf struggled to survive, and the Meistrells, who were superb divers, supplemented their income by taking on special diving projects. One project they did regularly—which still gives me the willies just thinking about it—was the inspection of the Edison power station's half-mile-long intake pipe. Their job was to check for obstructions and the general condition of the pipe—on the inside, mind you, not the outside! They entered the pipe from the land end with twin seventy-two-cubic-foot scuba tanks and their underwater camera and notepad. The water flow would then push them the half mile through the pipe to the deep end, where, so the plan went, they would emerge and come to the surface. If they had encountered a major obstruction, however, there would have been no way to reverse and go back because of the current pushing them through. All I can say is, they deserved a princely sum to do that kind of work.

The Meistrells also ran dive boat trips to Catalina Island. I was sometimes invited along if they didn't have a full load of paying customers; this gave me the opportunity to dive in places I couldn't reach in my little boat. One of the most spectacular and memorable sites we visited was the Farnsworth Bank seamount on the far side of the island. It wasn't really a bank, more like a rocky pinnacle jutting up from deep water to within sixty feet of the surface. The location on the seaward side of Catalina Island meant there was almost always a strong current or surge from the swells rolling in from the open Pacific Ocean.

The top of the seamount is quite small in area and it is covered with a forest of algae called sea palm (*Eisenia*), which thrives where there's a lot of water movement. Dropping down from the top you come across large clumps of the beautiful California purple hydrocoral (*Stylaster californica*). Although Farnsworth Bank is now a marine preserve, collecting this purple coral for public aquarium exhibits was permitted in the early years. We found the best coral heads at 130 feet and deeper. It became a dangerous challenge to carefully chisel them loose and bring them up to the boat, given the limited bottom time and air supply we had at that depth.

On one memorable dive in 1966, on a Sea World boat, we located the top of the pinnacle and dropped anchor on it. My dive buddy John Hart and I were first in the water off the swim step at the stern of the boat. There was a strong current running and we had to work hard just to swim the fifty feet from the stern to the bow, where we could grab the anchor line to descend. As a result, when we started down we were already quite out of breath.

The water was clear but we couldn't see the bottom, so we kept heading down the anchor line. It seemed like we had gone quite a distance, but still no end was in sight. Finally I could just make out the seafloor. I looked at my depth gauge; it said one hundred and seventy feet, and we still had another fifty feet or so to go. All of a sudden I found myself struggling to breathe as I was hit with a bad case of nitrogen narcosis, the intoxicating effect on the brain of nitrogen in the blood under pressure. My mind was a blur but functioned just enough for me to realize that the anchor must have slipped off the top of Farnsworth Bank and we were now drifting away over very deep water. Quickly we started back up the anchor line and made it to the boat. By then I had a splitting headache and was wiped out for the rest of the day.

Analyzing the situation later, I realized the strenuous swim to get to the anchor line had caused a major buildup of carbon dioxide in my system. To this was added the mind-numbing nitrogen narcosis, and I was in big trouble. This event was a vivid demonstration to me that deep dives when you are physically stressed can be dangerous. I was careful to make all future deep dives only when I was feeling physically and mentally relaxed. If I didn't feel good, I didn't dive.

THE MYSTERIOUS SUNKEN PIER

Just off the beach of Redondo at the south end of Santa Monica Bay is the head of a precipitous submarine canyon. With the nearest natural harbor being far away in San Diego, the city of Redondo Beach took advantage of the canyon's deep water and the partial protection of the bay to erect a pier that could handle ships carrying goods and people to the growing area of Los Angeles. In 1913, however, the construction of an eight-mile-long breakwater created the safe, protected waters of the new Los Angeles Harbor, and the Redondo Beach pier soon lapsed into disuse. Remnants of the old wooden pier were still there underwater in 1961, and seeing them brought to mind images of the bustling activity that must have gone on decades before.

Parallel rows of the stumps of the old wooden pilings stuck out of the sand. Near shore, where the wave action was greater, the stumps were either buried or barely protruding, but near the canyon head in forty feet of water they were six to ten feet high. Although the old pilings had been treated with creosote, the preservative had not penetrated to their centers, and the untreated inside of the wood pilings had been eaten away by shipworms. The surfaces of the hollow pilings were completely covered with beautiful colonies of several different color varieties of the strawberry sea anemone (*Corynactis californica*).

The beauty of the anemones, plus the hollow nature of the pilings, gave me an idea: if the pilings could be cut underwater with a handsaw, sections could be installed in an aquarium exhibit to show the richness of life under a wharf.

With a number of scuba tanks of air and the help of John Prescott's muscle, we cut four nice pilings covered with anemones and abalone jingle shell (*Pododesmus cepio*). To prevent the life-covered pilings from drying and dying, we draped them with wet cloths and quickly transported them back to Marineland. Installed in an aquarium and with an assortment of wharf fishes added, the pilings looked just like the real thing. In fact, they *were* the real thing! One drawback to the new exhibit was that even after decades underwater the creosote was still slowly seeping out of the wood where we had cut it. The addition of a continuous flow of fresh seawater kept the small amount of creosote from poisoning the animals.

On another dive near the old sunken pier, I was swimming along minding my own business when all of a sudden something grabbed my leg. I jumped and looked back to see a foot-long fish with its mouth firmly clamped on my wet suit–covered calf. I did a proverbial double take, and as soon as I stopped swimming the fish must have realized that perhaps it had bitten off more than it could chew. It let go of my leg, darted back down to the bottom, and ducked into a large wavy top shell. I recognized the fish as a sarcastic fringehead (*Neoclinus blanchardi*). I picked up the shell and popped it and the resident fish into my collecting bag to become part of the newly set-up pier piling exhibit. It turned out the shell contained developing eggs, which meant my attacker was a male fringehead standing guard over them. He certainly took his task seriously and was doing an admirable job of defending his future offspring against intruding giants like me.

Sarcastic fringeheads are remarkable-looking fish with big mouths and little tufts on top of their heads. The males have huge mouths that they use in dramatic displays with other males. Face-to-face, their mouths wide open, they display to each other by exposing the bright yellow lining of their mouths. Like the similar displays of French grunts (*Haemulon flavolineatum*) in the Caribbean, this mouth-to-mouth behavior settles mating and territorial disputes without actual fighting. I'm not sure how they determine who wins, whether it's the largest mouth or the brightest yellow lining, but somehow sarcastic fringeheads work out their social and territorial differences peacefully. Too bad we supposedly intelligent humans don't follow their example in settling our disputes.

RE-CREATING NATURE

The goal of aquariums is to show visitors the world beneath the surface of the sea. However, bringing that world into an aquarium intact is nearly impossible because the animal and plant growth is usually on rocks that weigh many tons and are virtually immovable. Nevertheless, I deeply wished to re-create the true underwater world in an aquarium setting. Eventually I came up with a scheme—but it involved some long-range planning.

Jerry Goldsmith and I gathered a pickup truck load of rocks from near the shore of Palos Verdes and drove them back to Marineland. Selecting from this pile, I carefully fitted them together so they lined the walls of an empty aquarium tank like a natural jigsaw puzzle. After making a sketch of all the pieces and how they interlocked, we took them out to a depth of forty feet on one of the biologically rich reefs in Marineland Cove.

The plan was to dive on them periodically and monitor the progress of developing growth. After two or three years, when the growth on the rocks resembled that on the natural rocks around them, we would collect and install them in the aquarium in exactly the configuration I had sketched out. It would be hard work, but I had a vision of the exhibit looking as close as we could get to the real underwater world. This ambitious plan was not to be, unfortunately, for I left Marineland shortly afterward. That means that somewhere out there on that reef, my carefully selected rocks must look perfect by now.

COLORFUL ROCKFISH

All fish are interesting to an ichthyologist. But the average aquarium visitor wants to see lots of colorful fishes, as well as exciting creatures like eels, octopuses, and sharks. It seems we humans derive pleasure from seeing bright colors, whether on fish, birds, butterflies, or flowers. That's not a problem if one is exhibiting creatures from the tropics, where there are hundreds of colorful fish to choose from, but here in California there are few colorful fish living in the shallow, diver-accessible waters.

One vivid California species, the brilliant-orange garibaldi (*Hypsipops rubicunda*), is brightly colored for a good reason. It's advertising the fact that it's a tough dude and if you're another garibaldi you'd better keep your distance from its territory. Crowding these fish in an aquarium will result in constant fighting and an unhealthy level of stress. A tank full of colorful garibaldi, therefore, doesn't work. One or two in a fairly large tank is about right; more than that means trouble.

Other colorful fish live along our Pacific Coast. They're in the rockfish family (Scorpaenidae), a family of about sixty-five species, and most are marked by shades and patterns of red. Even though they're one of the most common groups of fishes caught by both commercial and sportfishermen, the colorful ones are rarely seen in aquariums.

The most brightly colored rockfish live in moderately deep water. They have a large, gas-filled swimbladder that enables them to conserve energy by making them neutrally buoyant. That works out just fine as long as the fish stays pretty much at the same depth, but the large swimbladder becomes a major problem for the aquarium collector. When captured on hook and line in deep water and brought rapidly to the surface, the fish can die. The gas in the swimbladder obeys the laws of physics and expands as it encounters the lower water pressure near the surface. Unfortunately, these fish are not able to release the rapidly expanding gas very quickly. The result is that the swimbladder inflates to the point where it can actually force the stomach out of the fish's mouth. Other unpleasant and usually fatal results are popped eyes and a terminal case of the bends.

The few species of rockfish that live in shallow water are generally less colorful, and the deepwater species with their attractive shades of red come up either dead or dying when collected by conventional methods. This is fine for the dinner table, because rockfish are excellent eating, but not so good for the aquarium collector. (On the positive side, my profession is one of very few where you get to eat your failures.) Colorful rockfishes, though common lying on ice in the fish market, became a personal challenge for me, and I racked my brain trying to come up with a practical method that would give us live, healthy, and colorful rockfish for display in an aquarium.

I took on this rockfish challenge in 1962 at Marineland of the Pacific. My first, somewhat harebrained, scheme involved the Mark I decompression tank, which was intended to allow the fish to adjust slowly to the lower pressure at the surface. Actually, it was a home pressure cooker, which I modified so that either oxygen or air could be bubbled into it at about the same pressure found at the depth the fish was captured. A pressure regulator maintained the pressure and at the same time released excess gases. Over several hours the pressure was gradually reduced; then the cooker was opened up and, hopefully, the live, decompressed fishes taken out. We took the pressure cooker down with us when we dove, and as we collected the rockfish by hand net we put them in the pot and sealed the lid.

Logistically, the difficulties multiplied when Marineland's curator, John Prescott, built a large, heavy pressure tank—the Mark II decompression chamber. When it was full, a boom and winch were required to lift and lower the tank into the water, and there was no communication between diver and crane operator. Because the fish still had to be caught by hand, we were also limited to diving depths that were safe for us humans, and these were rarely deep enough to find the colorful species.

A WHALE OF AN ENCOUNTER

One of the early tests of John Prescott's Mark II fish decompression chamber took place in 1963 at a place called Eagle Reef on the lee side of Catalina Island. John and I were down at a depth of around eighty feet looking for suitable specimens to collect when all of a sudden we were startled by the presence of a huge animal. It was a gray whale (*Eschrichtius robustus*), swimming down toward us from the top of the reef! It circled around us, then went back up and over the reef. A few moments later it reappeared, curiously looking at us. It repeated this one more time, then vanished once and for all.

We'd never heard of anyone seeing a whale underwater, and we had no idea what it would do. Scared at first, we backed ourselves up against the rocks and watched as the whale circled. I realize now that it had never seen anything like us either and was just as intrigued by us as we were by it. Although the whale looked huge to us at the time, it was actually a young one of twenty feet or so. Perhaps its mother told it to stop fooling around with us and get back up with her.

About a year or two later I was lucky enough to run into gray whales in the water off Monterey, at a location known as the shale beds. My dive buddy, Dennis Sullivan, and I had run low on air and were reluctantly heading back to the surface and the boat. Relaxing, I was looking down toward the bottom as I slowly rose. The water was murky, as it most often is at the shale beds. All of a sudden I had a strange feeling of vertigo: for a moment it looked as if the entire bottom of the bay was moving sideways. I did a double take, then realized that what I was seeing was two gray whales passing ten or fifteen feet beneath me. It was a mother and calf, both rolled over and swimming on their sides, looking straight up at me.

Dennis had been looking in a totally different direction and missed them completely. Quite skeptical of my story, he was finally convinced—and jealous—when he saw the two whales blow once on their way toward Monterey Harbor.

These Rube Goldberg devices actually worked some of the time, but the major drawback was that there was no fresh seawater being pumped through; all it took was one fish to up-chuck its lunch or defecate and everything inside died. Once on board or on shore, the fish were at risk from the same dangers that deep-sea divers face. A fish can get the bends if the pressure is reduced too rapidly, or it can suffer from oxygen poisoning if the oxygen concentration is too high relative to the pressure. Eventually the whole method was abandoned, judged to be great in theory but no good in practice. It was frustrating at the time to concede defeat.

Three years later, biologist Dan Gotshall with the California Department of Fish and Game published a paper describing the use of a hypodermic needle to release the gases from the expanded swimbladders of rockfishes brought up from the deep. Now, this was more like it; the idea was so simple, with no complicated decompression equipment to contend with. It quickly became the universal method for dealing with expanding swimbladders of all kinds of fish, from rockfish to butterflyfish and even, as I would later have occasion to find out, flashlight fish from the Indian Ocean.

DESIGN FOR DINERS

Designing realistic exhibits has been an inspiring challenge throughout my career. I remember an especially unusual situation years ago when I applied at Sea World lessons I had learned during my Marineland days. The park was building a fancy restaurant on Mission Bay, and of course, as part of Sea World, it had to include striking aquariums. I designed four totally different tanks: two behind the bar, one in the cocktail lounge, and a large one that separated two banquet rooms.

Consider the challenge of installing functional aquariums in an operating restaurant. Somehow it worked out, but neither the aquarium nor the restaurant staffs were too happy with their work spaces. The occasional saltwater flood on the floor irked the restaurant staff for some reason, while the bus boys feeding the fish leftover cheese and dinner rolls irritated us aquarists.

The cocktail lounge housed a six-thousand-gallon aquarium for fishes from the Sea of Cortez. The tank, high and narrow from front to back, was too deep to effectively clean the glass with a pole from the top, so to do a good job required diving in a very cramped space. Although the tank was attractive, space constraints made it barely wide enough for a diver to fit between the front glass and the back rockwork. The banquet room's ten-thousand-gallon aquarium exhibited southern California game fishes. For it we collected a nice school of yellowtail, some leopard sharks, sheephead, and a couple of large black sea bass.

The two tanks behind the bar were designed to show striking contrasts in color. One had dark volcanic rockwork with pure-white powder puff sea anemones and bright red rockfishes. The other tank had large red Tealia sea anemones with contrasting silvery surfperch. Those two tanks were quite attractive but led to some interesting collecting adventures. In particular, we had to collect enough colorful red rockfish to create the color contrast with the white sea anemones.

Milt Shedd, chairman of the board of Sea World and an expert angler, kindly offered the use of his sixty-seven-foot sportfishing boat, appropriately named *Sea World.* Several times Milt took us to Santa Catalina and San Clemente Islands to collect rockfish.

We'd developed a technique by which, theoretically at least, we could successfully collect live rockfish from as deep as three hundred feet. We knew from past experience that fish brought to the surface from that depth wouldn't survive even when their swimbladders were deflated with a hypodermic needle at the surface. The new plan was to catch the fish on a multihook, squid-baited drop line. As soon as the line was lowered to just above the bottom, we would get bites. When we judged that several fish were hooked, we would reel it up until a mark on the line indicated that the hooks and fish were hanging at about one hundred feet. We would then dive down and deflate their swimbladders at the hundred-foot stop, and then again at another shallower stop on the way up to the boat.

Bob Kiwala, collector for the Scripps Institute of Oceanography, had designed and constructed a nice heavy rockfish line on a large and easy-to-use spool. We didn't like monofilament fishing line because it's difficult to see underwater and we were nervous diving next to an almost invisible line with numerous hooks moving up and down with every swell that raised and lowered the boat above. We didn't relish the thought of being snagged by one of those hooks a hundred feet down.

Our new system actually worked and we were able to collect several species of beautiful rockfish never before displayed in aquariums, including the colorful and highly prized rosy (*Sebastes rosaceus*), vermilion (*S. miniatus*), and starry (*S. constellatus*), as well as the green-spotted rockfish (*S. chlorostictus*), which to our knowledge had never before been collected alive. Later we received many requests from other aquariums to trade for these fish. Little did they know the time, effort, and risk that went into collecting them.

The water at the Channel Islands, which include Catalina and San Clemente, is generally very clear. This is the habitat of the pelagic blue shark (*Prionace glauca*). On more than one occasion blue sharks were attracted to the drop line with its string of struggling rockfish. Although they weren't a serious threat to us, we did get a little nervous when a six-foot blue was chomping our hard-earned fish from the line while we were trying to needle another one just two hooks away.

One rockfish-collecting trip I will never forget took us to the lee side of San Clemente Island, a hundred miles seaward of San Diego. We'd made one set and brought some nice fish on board, and then moved on to another location. The baited line had been dropped, fish were hooked, and we had brought the line up to a depth of 80–100 feet. Just as Sea World aquarist Kym Murphy and I jumped in the water to head down to needle our catch, the strap on my old tank backpack broke. I signaled to Kym that I was going back to the boat to get another pack and would join him at the fish.

Back on board and frustrated by my very old, now broken backpack, I threw it overboard and grabbed another. I was just getting ready to jump in when Kym broke the surface and yelled, "Shark!" He had seen my backpack sinking into the depths and, assuming I had accidentally dropped it, swam down to retrieve it. At the same time, a very large, gray-colored shark came up out of the depths heading toward the same pack, possibly attracted by the shiny stainless steel band, or perhaps by Kym himself. I suspect Kym broke the swimming speed record from that point to the boat above.

We're not sure what kind of shark it was, other than a very big one; it was definitely not one of our familiar, slender blue sharks. The most likely shark that fits Kym's brief description would be the great white (*Carcharodon carcharias*) or possibly a dusky (*Carcharhinus obscurus*), but for some reason he didn't stick around long enough to make a positive identification, as a truly dedicated ichthyologist should. After that incident, in any case, we thought it prudent to move to a spot far, far away to continue our collecting.

4

THE ROAD TO GONZAGA BAY

I FIRST MET DAVID ANDERSON at a Marine Aquarium Society meeting in Los Angeles. Dave and his wife, Carolyn, ran an aquarium supply business out of their home in southern California. They had tanks of freshwater fishes and plants throughout their house, garage, and backyard and were breeding fishes that they then delivered to pet shops throughout the area.

Dave told me he wanted to begin handling marine fishes as well as freshwater species, and with that in mind he'd made a reconnaissance trip down to the northern end of the Gulf of California. He had a 1957 rear-engine Chevrolet Corvair and had driven it to a place called San Luis Gonzaga, a hundred miles past the end of the paved road at San Felipe. Since Dave was not a scuba diver, he asked if I would be interested in going with him on his next trip to see if there might be some fishes suitable to sell. It seemed like an exciting adventure; I said I'd do it.

I packed my dive gear, two full scuba tanks, and my collecting hand net, while Dave took care of the camping gear, fish-shipping boxes, bags, and oxygen, on the chance we caught fish we wanted to bring back. He also took a variety of tools in case we ran into car trouble miles from civilization. Leaving L.A. early one morning, we crossed the border into Mexico at Mexicali, gassed up the car in San Felipe, and headed out on the one-lane dirt road to Puertecitos fifty miles south.

That stretch of road is crossed by numerous soft, sandy arroyos, and I was as impressed with the way the little car went through sand as I was with Dave's driving skills. When we arrived at the tiny fishing settlement of Puertecitos, Dave said he needed a break. He proceeded to lie down on the ground and fall fast asleep. Twenty minutes later he was up and ready to go. Clearly this was an energetic and driven man. We took off again for San Luis Gonzaga, another fifty miles south.

The terrain changed considerably and the road began to wind up over steep, rocky hills. The roadbed was now jagged rock cut by frequent eroded gullies. The car had hit bottom many times on the sandy road to Puertecitos, and now it was doing that on the rough road of hard rock. Taking a run to get up an exceptionally steep hill, we hit bottom really hard; this time, we feared, we'd done some damage.

Peering under the low Corvair we saw a trickle of fluid coming from the pan of the automatic transaxle—definitely not a good sight in such an out-of-the-way place. We couldn't do anything on that steep, narrow road, so Dave started the car, made it over the hill to a wider spot in the road, and pulled over. Dave managed to get the car jacked up enough so he could slither underneath to catch the leaking transmission fluid in a can and to check out the size of the hole. He managed to save about a quart of the fluid trickling out of a small crack in the metal pan.

I was amazed at his calm attitude in a situation that to me seemed quite serious. He approached it as if it were simply one more problem that needed to be solved so we could be on our way again. With the set of tools he'd brought he dropped the pan, cleaned it thoroughly with gasoline, and, using a propane torch, solder, and a piece cut from a beer can, he soldered a patch over the crack. That done, he reinstalled the pan, poured the saved fluid in plus an extra quart he had brought along, and topped it up with some engine oil thinned with a little gasoline.

Once again we started out, and the Corvair seemed to run just fine. Arriving in San Luis Gonzaga safely, we set up camp, had a quick meal of canned stew and fresh Mexican rolls, and turned in for the night.

Bahía de San Luis Gonzaga is an almost totally enclosed, well-protected bay with a narrow entrance out to the Gulf of California. This remote and desolate place was home to the fishing family of seventy-year-old Papa and Mama Fernandez for many years. The Fernandezes, who form a colorful part of Gonzaga Bay's local legend, were most gracious to us as we went about our business.

DISCOVERING THE CORTEZ ANGELFISH

The next morning we decided to snorkel along the shore of the entrance channel to see what was there. I put a scuba tank on in case I spotted something down deeper and needed air to check it out or to catch it. The visibility was poor: we could see only about ten feet, so to inspect anything deeper than that we had to dive holding our breath. We had gone only a short way when I glimpsed a flash of blue and yellow dart under a ledge. Snorkeling down to investigate, I was thrilled to see a beautiful two-inch juvenile angelfish. Using air from the scuba tank so I could stay down, I chased it from under its ledge with my hand and it shot right into my strategically placed hand net.

Back at the surface Dave and I got a closer look. It had a velvety black body with four or five crescent-shaped bright yellow bars and a touch of metallic blue on its dorsal fin. To us, who were familiar with the generally drab colors of California fish, it was spectacular. Dave was as excited as a kid who had just opened his first Christmas present.

He wanted more. We set off again, making our way slowly out toward the Gulf. Quickly Dave spotted another one, and I went down and netted it. The water was gradually getting deeper, and I needed to take a few breaths from the scuba tank to get down to where I might see more angels. By the time we reached the opening to the Gulf we had twelve angels in the bags and it was time to head back in.

Out near the point I had seen several pairs of very large grayish-colored angelfishes marked with vertical yellow bars. I knew from my UCLA ichthyology classes that young angelfishes are often radically different in appearance from the adults, and I wondered if what we had caught could be the juveniles of these adults.

In 1961 there were no identification books on the fishes or invertebrates of the Sea of Cortez, but I had brought along a checklist of fishes

A pair of adult Cortez angelfish. (Photo by author)

of the Gulf compiled by my ichthyology professor, Boyd Walker, and Ken Norris. Only one species of angelfish was indicated from this area of the Sea of Cortez: *Pomacanthus zonipectus.* Surprisingly, though, this beguiling little fish with its striking appearance had no common name—an injustice that Dave and I pondered over a couple of beers. We decided that the name Cortez angelfish was perfect for introducing this fish to the aquarium world. Not only did it have a nice romantic ring to it, but it was also geographically appropriate. As more and more aquarists and collectors began to travel to the Gulf region to collect their own angels, the word spread and the common name Dave and I coined became accepted throughout the aquarium world. Now when I hear someone talk about the Cortez angelfish, I chuckle remembering how the fish received its popular name.

The next day was as successful as the first. By the time we had to head back to the States I had sucked my two tanks of air dry and we had twenty-two angels ranging in size from one inch to three inches. Dave now had some fish that would sell very well in the aquarium shops.

That last morning we piled all our gear and the precious Cortez angels into the little Corvair and headed back out on the road from hell, to Puertecitos, San Felipe, and home. This time we had no trouble, even though the tires were taking a frightful beating from the rocks and the heavily loaded car was hitting bottom on a regular basis.

Back in Los Angeles Dave sold all the angels in record time and got orders for more. At this point he realized that if he was to continue making regular trips to San Luis Gonzaga he had to invest in a better vehicle. The Corvair was one of the lowest-riding cars on the market and certainly poorly suited to the kind of rough road we had just traveled. So Dave, obviously serious about making more trips to Gonzaga Bay, bought a brand-new 1962 Chevy half-ton pickup truck.

In addition to a better vehicle, we also knew that some kind of diving air would be needed. It didn't make sense to haul a bunch of full scuba tanks down—even if we had them, which we didn't—and then haul the empty tanks back. Since the little angelfish we were after were in shallow water, I suggested that we rig up a low-pressure air compressor that could supply me with continuous air to the second stage of my regulator—an arrangement known as a hookah after the Turkish hoselike apparatus used for smoking tobacco and hashish.

We bought an inexpensive diaphragm air compressor designed for spray painting—powered by a gasoline engine—and fifty feet of nylon-reinforced garden hose to connect to the second stage of my two-hose regulator, which was tied to a plywood board I would strap to my back. The engine and compressor were mounted on plywood lashed to a large truck inner tube. The plan was for me to tow the floating hose and air compressor along as I cruised the bottom looking for fish. It worked great out of the water, and we planned our next trip down to San Luis Gonzaga.

The new truck was a vast improvement over the Corvair, though it did get stuck a couple of times in the soft, sandy arroyos on the way south. The floating air supply worked just as we hoped it would; all we had to do was keep it supplied with gas, and I could dive as long as I wanted without worrying about using up precious scuba tank air. This trip was even more successful than the first: we came back with thirty-five Cortez angels, which again sold like hotcakes.

As a full-time employee at Marineland, my free time for trips like these was limited. My family, too, had grown with the birth in 1960 of our second daughter, Amy. What with my job, my wife, our two little girls, and my own weekend diving forays, Dave Anderson realized he needed to arrange for a new partner if he wanted to continue his commercial collecting trips. He also figured he should learn to dive so he could do the work himself.

He did just that: he learned to dive, bought his own dive gear, and found other people who were free to accompany him on a number of four-day collecting trips. Dave and I made only one more trip to Gonzaga Bay, this one with Betty and our two young daughters.

The story of my friendship with Dave Anderson has a very sad postscript. By 1966, when our family had moved to San Diego and I was working at Sea World, we saw Dave infrequently. That year, while I was on a collecting trip to Cabo San Lucas, the news broke that Dave's body had been found in a shallow grave near a dirt road in Baja. He had apparently been murdered.

You read about murders every day in the newspapers but never imagine they could involve someone you know. How ironic that Dave Anderson, who was so capable of surviving life-threatening situations in the wilds of Baja California, would meet such an unhappy end.

5

THE STEINHART AQUARIUM

AN OFFER FROM THE STEINHART AQUARIUM in San Francisco prompted a major move for me and my family in 1962. Joining the staff of one of the most prestigious aquariums in the United States, I knew, would be a great opportunity for me to learn about aquatic animals from all over the world. Not only that, but the Steinhart was in the middle of its first complete renovation since it opened in 1923, so the timing couldn't have been better.

This relocation came about after Dr. Earl Herald, director of the Steinhart, had visited Marineland. I wasn't aware of his visit and didn't meet him at the time, but he must have been impressed with the exhibits he saw, since he wrote and offered me the job of assistant curator. Of course, I accepted.

And so, with all our worldly belongings crammed into a sixteen-foot U-Haul trailer that dwarfed our little 1950 Ford "woody" station wagon, we headed north. What a sight we must have been! Our entourage included my wife, Betty, our preschool-age daughters, Eve and Amy, two cats, a tortoise, twenty or so marine creatures from my home aquarium, and me. The trip along the coast took three days, with two nights spent in cheap motels.

Arriving in San Francisco, we were dismayed to find that housing prices in the city were much higher than the $85 a month we had been paying in Redondo Beach for a small house one block from the beach. We rented a flat on a typical steep San Francisco hill and settled in.

Steinhart Aquarium is part of the California Academy of Sciences, founded in 1853. Originally located in downtown San Francisco, the Academy together with much of its collection was lost in the earthquake and fires of 1906. It was rebuilt in Golden Gate Park, with famed ichthyologist Dr. Barton Warren Evermann as its president. The Academy has had a special interest in both the scientific study and display of fishes ever since. In 1916 Ignatz Steinhart, a wealthy banker and philanthropist, endowed $250,000 for the construction of a major aquarium in San Francisco, stipulating that it be located in Golden Gate Park and be operated by the Academy. World War I delayed the construction of the aquarium, but it finally opened in 1923 with an estimated record twenty-five thousand visitors the first day.

When I arrived at the Steinhart, the original concrete tanks were being demolished and replaced with new tanks reinforced with corrosion-resistant Monel bars, and newly developed PVC plastic pipe was being installed in place of the old piping. A marine mammal tank was being added for dolphins and seals in addition to three new sections for small exhibit tanks.

I worked under Earl Herald and curator Bob Dempster. As assistant curator, my primary responsibility was the health and welfare of the animals. This became quite tricky during the reconstruction because most of the aquarium's fish, reptiles, and amphibians still needed to be cared for even while their tanks were being demolished and rebuilt. Local fishes that would be relatively easy to collect again were released back where they had come from; this created space to hold others while their new homes were being built. It was a major fish juggling act as fish were moved from gallery to gallery in advance of the construction crew.

The eight alligators, each six to eight feet long, were temporarily housed in a well-fenced swimming pool at a pool supply company across the Golden Gate Bridge. This involved making trips three times a week across the bridge lugging buckets of fish to feed them. I was always surprised how fast and agile an alligator can be after basking in the hot summer sun of Marin County.

When all the tanks were completed, the temperate fishes that had been temporarily held in the new marine mammal tank were moved to their new permanent quarters. The mammal tank was cleaned and refilled and made ready for its new occupants, two Pacific white-sided dolphins from Monterey Bay and two rescued harbor seal pups (*Phoca vitulina*).

MONTEREY WEEKENDS

Before too long, my cousins Norman and Wilkie Powell trailered my boat up from Redondo Beach to Monterey, two hours south of San Francisco. I found a place to store it in the abandoned Monterey Canning Company building, right next to Ed "Doc" Ricketts's old Western Biological Lab. I could peer through cracks in the cannery wall and see the tanks where Ricketts had kept his marine animals.

On weekends I drove down to Monterey from San Francisco, launched my boat in the harbor, and went diving. At the time, Steinhart Aquarium didn't have a collecting boat or even much in the way of staff who could dive. I therefore began to collect animals for the aquarium on my days off. Some of the Academy staff, including Herald's secretary, Phyllis Ensrud, and aquarist Lloyd Gomez, were divers and would sometimes come with me. In spite of not having a truck with holding tanks, I was able to bring colorful invertebrates and fishes back to San Francisco in our 1957 Chevy station wagon. Among my prizes were beautiful, then-undescribed Tealia sea anemones (*Urticina piscivora*), with their red body and pure-white tentacles, and colorful painted greenlings (*Oxylebius pictus*). They, and many others, made wonderful additions to Steinhart's exhibits of local animals.

A number of times a buddy diver and I would camp overnight with our sleeping bags in what later became Monterey's Veteran's Memorial Park so we could do some night diving. The night diving I had done for lobsters in southern California made me curious about the creatures I would see in the colder waters of central California. I learned that you don't have to travel to exotic places to see new animals or

Ed Ricketts *(right)* and Richie Lovejoy examining a bat ray specimen at the Pacific Biological Laboratories, April 1939. (Pat Hathaway Photo Collection, Monterey)

THE LEGEND OF DOC RICKETTS

Ed "Doc" Ricketts was a pioneering marine biologist who operated a small biological supply house on Cannery Row from 1923 until he died in 1948, when his car was hit by a train only a few blocks from his laboratory. Ricketts made a modest living collecting, preserving, and selling biological study specimens to schools and universities. The selection of animals used by the universities for anatomy classes was broad: frogs, cats, spiny dogfish sharks, and a wide range of invertebrates including marine worms, sea stars, brittle stars, and mollusks such as octopus, giant geoduck clams, and gastropod snails. His travels to collect the specimens covered the entire Pacific Coast.

Ricketts was not one to adapt to the structured way of thinking in the academic world. After all, his keen observations of the way animals and plants have adapted to life in the harsh and competitive world of the intertidal gave him new insights into the way things worked.

The physical environments along the shores of the West Coast, he noted, differ widely in types of bottom and exposure to waves and to air during low tides. These differences have led to the evolution of invertebrates that are adapted to particular habitats and physical conditions. Eventually Ricketts was talked into writing the now classic book *Between Pacific Tides,* which clearly laid out the zonation of life in the intertidal.

Ricketts had a fascination with all life. His charismatic Bohemian nature attracted a number of friends of both sexes, among them writer John Steinbeck. Ricketts and Steinbeck formed a close friendship that culminated in 1940 with a scientific expedition to Baja California and the Gulf of California. The success of Steinbeck's book *The Grapes of Wrath* enabled him to charter a boat, the *Western Flyer,* and fund the expedition.

The narrative part of the jointly authored book *The Sea of Cortez,* written by Steinbeck, is filled with the philosophy of Ed Ricketts. In addition, although Steinbeck took certain literary liberties, the character of Doc in the books *Cannery Row* and *Sweet Thursday* is clearly patterned after his friend Ed Ricketts.

Cannery Row as Ricketts knew it and the way it now exists as a major tourist area on the California coast represents two different worlds. However, his laboratory is still there, almost untouched. To many people, that lab is a memorial to those rough-and-tumble days; you can almost imagine Ricketts preparing to catch the early-morning low tide to make a collection and to wonder at the diversity of life along the shore.

---

behaviors, or even, as I discovered, to find a totally new species. You simply have to dive in the same old place, but at an hour when sensible human beings are engaged in more normal activities—like sleeping. We made our dives close to midnight, and we saw nocturnal creatures never visible during daylight hours.

The Monterey Coast Guard breakwater is a mass of granite boulders piled over fifty feet high from the bottom. This structure creates myriad homes for animals that shun the light of day, emerging only in the dark of night to forage for food. The beautiful but highly secretive fish the red brotula (*Brosmophysis marginata*), for example, could be seen close to the bottom of the rock pile, its bright red color visible only when illuminated by our bright underwater lights.

You wouldn't believe it possible to encounter an unknown fish in a frequently dived area like Monterey, yet on a number of night dives I collected specimens of what turned out to be a genus of fish totally new to North America. Described by the late ichthyologist Bill Follett of the California Academy of Sciences and me, this interesting-looking, eight-inch-long fish was given the name masked prickleback, or, in scientific parlance, *Ernogrammus walkeri,* after Boyd Walker, my UCLA professor of ichthyology. Unfortunately, all attempts to display it in an aquarium failed. Being strictly nocturnal, the poor fish hated light and would try to hide under any object it could find. If there was one little rock in its tank, it would wiggle under it and be out of sight. I suppose I could have taken everything out of its exhibit, but I felt that would be cruel and no doubt would have traumatized its dark-loving psyche. Training it to wear tiny sunglasses would have been a challenge.

6

SAN FRANCISCO AND THE SEA OF CORTEZ

THE CALIFORNIA ACADEMY OF SCIENCES was a stimulating, rewarding place to work. Just its location in the middle of beautiful Golden Gate Park was, to me, a perk. Often Betty brought Eve and Amy and we would sit in Shakespeare's Garden to eat our picnic lunch and feed the squirrels. We also made good friends at the Academy. The best were Dusty Chivers and his wife, Lynn, who with their son, Dana, remained our close friends until Dusty and Lynn died.

Dusty worked in the Invertebrate Zoology Department and loved his work. Over the years he developed a broad knowledge of the biology of this huge, diverse group of animals, especially those of San Francisco Bay and central California. Before we met, Dusty and his buddies were among the first scuba divers to explore the waters around Point Lobos and the precipitous Carmel submarine canyon that comes within mere yards of Monastery Beach, just south of Carmel. They made insanely deep dives—to over two hundred feet—somehow managing not to kill themselves from nitrogen narcosis, the bends, or hypothermia. In the days before wet suits they would dive in long underwear and try to warm up afterward by building a bonfire on the beach.

The highlight of my work at the Academy was participating in the June 1964 Scientific Expedition to the Sea of Cortez. The expedition was led by Dr. George Lindsay, director of the Academy, who had done considerable work on the cactuses of Baja California and the islands of the Gulf of California. Participants were selected on the basis of their knowledge in a particular specialty of biology. The team included, among others, Dusty Chivers and Alejandro Villalobos, invertebrate biology; Dr. Robert Orr, mammals; Ray Bandar, reptiles; Dr. Ira Wiggins, botany; Chris Parrish, scorpions; and Dr. Earl Herald, ichthyology. Dusty and I were the diving biologists, and George Tsegeletos from Marin Divers was hired as the underwater photographer. The expedition was funded by Academy supporter Roy Marquardt, founder of the Marquardt Corporation, manufacturer of rocket engines. Roy's teenage son Bruce went along with us.

EXPEDITION TO THE GULF

That June we flew from the Tijuana airport to Loreto in a twelve-passenger 1939 Lockheed Lodestar piloted by Francisco Muñoz. (This was the same type of plane Amelia Earhart flew in her ill-fated attempt to fly around the world.) At Loreto we boarded the converted landing craft *Marisla,* owned by American expatriate Dick Adcock and his Mexican wife, Marylou. It was equipped as a dive boat, complete with scuba tanks, dive weights, and an air compressor for filling tanks.

In the ten days of the expedition we covered almost all of the Gulf islands from Loreto south, on land and beneath the sea. It was my first time diving in tropical waters, and I was fascinated by the totally different and incredibly rich marine life of the Gulf of California. At Isla Las Animas, a remote rock pinnacle that juts up out of deep water, giant groupers (*Epinephelus itajara*) followed me around as I collected and photographed. It was as though they were wondering what this strange bubble-blowing creature was doing at their island. Here we had glimpses of the legendary hammerhead sharks (*Sphyrna lewini*), which congregate by the hundreds at certain times of the year around rocky pinnacles like Isla Las Animas and submarine seamounts like El Bajo. This was my first time to see colonies of wonderful garden eels waving in the current as they fed on tiny drifting zooplankton.

We'd set up holding tanks on board the *Marisla* to hold the fishes and invertebrates that would later be flown back to Steinhart Aquarium. By the end of the trip we had a fine collection of colorful butterflyfishes, angelfishes, squirrelfishes, wrasses, and assorted invertebrates. I had also managed to collect the ugly but very personable finespotted jawfish (*Opistognathus punctatus*). While I concentrated on the underwater life, the dry-land members of the expedition were equally focused on everything from scorpions and snakes to desert plants and birds. Discoveries of special interest were shared with all on board, and we learned a lot from one another.

The Sea of Cortez is home to a species of fish-eating bat, *Pizonyx vivesi,* several specimens of which Bob Orr and Ray Bandar collected from a crevice in the rocks on Isla San Francisco. This remarkable animal swoops low dragging its clawed feet in the glassy-calm water to snag small fish. It's fascinating to ponder how such a behavior might have originated among these flying mammals' insect-eating ancestors. Did one bat with exceptionally long claws one day swoop especially low over the water and "accidentally" snag a small fish? Did it then try again, succeed, thrive, and pass its genes on to its offspring? Did other bats see and copy this behavior? It's interesting and also fun to speculate on how such highly specialized behaviors as this begin.

The Gulf island of Santa Catalina in particular was a fascinating place both above and below the water. This island was home to the world's only rattleless rattlesnake. Did the absence of land predators eliminate the survival advantage of having a warning rattle? Most likely. What advantage, after all, is a complicated device like a rattle if there is nothing to frighten away? It's generally true in the biological world that a structure that no longer provides a survival advantage will tend to degenerate over many generations. The remnant hind legs of some snakes are an example of this degeneration.

As Charles Darwin so astutely observed on the Galápagos Islands, evolution can go in radically different directions when animals are placed in an environment quite different from that in which they originally evolved. Animals and plants can change rapidly under the pressures of natural selection in the new environment and also from what is called genetic drift. Genetic drift occurs when a new population is founded by a few colonizers, or in some cases even a single fertilized female.

These few individuals contain by sheer chance a fraction of the total genetic diversity of the mainland gene pool. The population that ultimately arises from the colonizers will be significantly different from the mainland population and will tend to diverge even further through the process of natural selection.

Another remarkable animal found on Isla Santa Catalina is the giant plant-eating chuckawalla lizard (*Sauromalus varius*). Throughout mainland California and Baja California chuckawallas are dark gray, a color that's excellent camouflage among the granite rock outcrops they inhabit. On Isla Santa Catalina, however, they're marked with colorful blotches of red and yellow over their gray body. It's hard to imagine what the survival advantage could be of such coloration, but perhaps, as with the rattlesnake, it's the absence of large predators that allows these gaudy lizards to survive just as well as their inconspicuous gray cousins.

The trip was filled with memorable experiences for everyone, most of them pleasant. However, a not-so-pleasant event involved a short-spined pinkish sea urchin (*Toxopneustes rosacea*) that I brought back to the boat to show the nondivers. I passed the urchin up to someone on deck, who then passed it on to Pedro Muñoz (Francisco's fourteen-year-old son). As he held it in the palm of his bare hand he let out a yelp, dropped the animal, and fell to the deck writhing in pain. When we examined his hand we saw three tiny red punctures on his palm. All species of sea urchins have small three-jawed pinching organs called pedicillaria between their spines, which serve to keep the larvae of invertebrates like barnacles from setting up housekeeping on the urchin's shell. This flower sea urchin, it turns out, has unusually large pedicillaria that contain a painful toxin that must serve to keep urchin predators like triggerfish from even thinking about feeding on one more than once. Instant pain is an excellent reason to leave them alone.

The intense experience of the ten-day expedition finally came to an end. Hundreds of preserved specimens of all types were unloaded in La Paz to be shipped back to the Academy. The more difficult task was to get the live fishes and invertebrates back to Steinhart Aquarium as quickly as possible. The first leg of the return trip was in a small private plane from La Paz across the border into the United States, where they were transferred to a commercial flight to San Francisco. This was my first experience in shipping aquatic animals by air in plastic bags and Styrofoam boxes. It would also be my first lesson in how not to do it.

Concerned about the length of time the animals would be in the bags, I filled the bags with as much oxygen as possible before sealing them. This was a big mistake. The unpressurized plane flew at an altitude of several thousand feet, where the air pressure was much lower than at sea level. The laws of physics being what they are, the oxygen expanded and many of the bags with their precious live cargo burst and collapsed. A number of fish died from lack of oxygen because of this error. It was heartbreaking to lose these beautiful animals, and it was a mistake I never made again.

Strong bonds of friendship resulted from the sharing of adventures in the wilds. When I recall this expedition I have especially fond memories of Dusty Chivers. Plagued by alcoholism in his middle years, our friend Dusty eventually managed to lick it and become a positive force in the San Francisco Alcoholics Anonymous. The memorial at the Academy following his death from cancer in 1995 was attended by people from many walks of life, including quite a number who owed their sobriety—and literally their lives—to Dusty. It was very moving to see how his life had touched others and how he would be missed by so many.

MAKING WAVES

Working with Earl Herald was a bit of a challenge. He had strong opinions, and rarely could he be swayed to believe differently. As the reconstruction of the aquarium progressed, we reached a point where we needed to figure out what kind of background should go into each newly poured concrete tank. These were the days before naturalistic fiberglass or gunite aquarium interiors, and so it was decided to line the tank walls with various types of natural rock obtained from decorative rock yards.

Herald was a fish person and firmly believed that the purpose of an aquarium was to display fish—the more fish the better. When I was put in charge of selecting and supervising the installation of the rock on the tank walls, Herald told me he didn't want any rock protruding more than three inches because it would reduce the space that could be occupied by fish. Needless to say, as a diver I wasn't too happy with tanks that looked like someone's garden wall.

Herald came up to the exhibit floor every day to see how the tank decorating was going. At one point in the process, though, he left town to attend a conference. Now was my chance. Knowing that many species of fish are crevice or shelter dwellers, I instructed the rock masons to create some caves, ledges, and overhangs that could be utilized by fishes that would feel uncomfortable in a wide-open tank.

When Herald returned from the conference he immediately came to see how the rockwork was progressing. He was furious! Of course, by then the mortar had set up hard and to remove it and redo it would have put the project behind schedule and over budget, so my attempt to create a little variety in the habitat was safe. Although I was on Herald's blacklist from then on, I like to think that some of the fish appreciated what I'd done.

I wasn't totally frustrated in my attempts to create more naturalistic exhibits, however. Included in the plans of the aquarium renovation was a new intertidal exhibit that would display plants and animals living in the tide pools along the central California coast. As Ed Ricketts understood so well, there are two vital forces shaping life in the tide pools: waves and tides. Certainly, a tide pool without waves—the most visually striking of these two forces—is just a motionless pool.

As at Marineland, I was intrigued with the idea of seeing waves crash into an exhibit like this, and I set about figuring out how to make it happen. I started experimenting on a small scale. I took a cylindrical reservoir (actually a plastic garbage can) and, in the bottom, made a large round hole. I then blocked the hole with a plastic beach ball. According to my plan, when the water in the reservoir rose to a certain height, a float would lift the ball off the hole and suddenly release all the water from the reservoir. This sounded very straightforward, and a consultation with an engineer confirmed I was on the right track.

I put the contraption together and started running water into the reservoir, watching anxiously as the water level rose, lifting the float.

The line between the float and the ball tightened, but instead of being jerked off the hole, the ball lifted up a quarter of an inch and then snapped back down. As the water flowed into the reservoir, the ball continued to bounce up and down, letting a little bit of water out with each bounce. All of a sudden, memories of my UCLA physics classes came back to me, and the problem became clear. What I had here was a perfect illustration of Bernoulli's principle, where high velocity creates low pressure—the same principle that allows airplanes to fly. When the water rushed out below the ball, it created low pressure, which promptly pulled the ball back down. As soon as I realized my idea would never work—unless, that is, I could repeal the laws of physics—I junked it and began work on another.

My new plan was to create an automatic siphon. The beauty of this design was that it had no moving parts—other than a supply of water. I built a small-scale model out of an old bucket. Using heat, I bent a piece of PVC pipe into a U shape, with one side of the U much longer than the other. Drilling a hole near the top of the bucket's wall, I installed and sealed the U-shaped pipe so that the short arm was just off the bottom of the bucket and the long arm extended well below the bucket on the outside. I started filling the bucket, and as it filled, the short end of the pipe also filled. As soon as the water reached the top of the pipe, it started spilling over into the long side. This rapidly falling water carried the air out with it, until quickly the entire pipe was full of water. At that point the pipe became a siphon and the bucket emptied with a rush. Once it was empty, the siphon stopped and the process started over again. It was so simple!

Now came the scary part: scaling it up in size for the six-hundred-gallon tide pool exhibit. This was going to cost the aquarium a lot of bucks, and I was still not one hundred percent sure it was going to work. First, a two-hundred-gallon fiberglass reservoir was mounted on a steel platform bolted to the wall fifteen feet above the tide pool. Then a twenty-foot-long piece of six-inch-diameter PVC pipe was heated in a giant oven and curved into a big, lopsided U shape. It was installed through the side of the reservoir as in my mock-up, with the end of the long arm just penetrating the water surface in the tide pool below.

I was very nervous when the pump was turned on to begin filling

Daughters Eve and Amy admiring Steinhart Aquarium's new Pacific white-sided dolphins. (Photo by author)

the reservoir. Would it work? I breathed a huge sigh of relief when over a thousand pounds of seawater came rushing down into the tide pool, turning it white with foam. In a few seconds the water cleared. Not only was it visually dramatic, but the sea anemones and tide pool fishes loved it. This wave machine, first turned on in 1964, has been surging into the tide pool every fifty seconds ever since, and as far as I know may still be going strong.

The day finally came when the renovation was finished and the aquarium reopened amid much fanfare. The crowds were incredible. I later heard that we had two and a half million visitors the first year. The new marine mammal exhibit with the two Pacific white-sided dolphins and the two harbor seals was a tremendous crowd pleaser, and their feeding times were always packed. I still enjoy mingling incognito among the visitors and listening to their exclamations of amazement at what they are seeing. It's very rewarding to get such enthusiastic responses to your work. It's a good feeling.

I first met Bob Kiwala shortly after the aquarium reopened. He'd been working temporarily up at the University of California's Bodega Marine Station and had started at Steinhart as an aquarist working with assorted species of fishes. A short time after Bob arrived, Earl Herald came up with the idea that everyone should be able to do everything. Although Bob's interests, training, and expertise were in marine biology, he was ordered by Herald to work with the reptiles.

Herald's specialty was ichthyology, but he had a strong interest in all cold-blooded vertebrates, and as a result Steinhart Aquarium had an extensive collection of reptiles and amphibians as well as fish. These were all under the direction of one of the herpetologists, who shall go unnamed. I wasn't able to learn much about the herpetologist's professional background, but I did know that he had a fascination with venomous snakes. He loved to show off by handling deadly ones; the more dangerous they were, the happier he was.

This herpetologist had accumulated quite an extensive collection of snakes from around the world. He had more species of rattlesnakes than I knew existed, as well as cottonmouths, Mexican cantils, and coral snakes from North America—but they were the tame ones. He had regular cobras, golden cobras, a twelve-foot-long king cobra (which ate only other snakes), and the charming spitting cobra, which, aiming for your eyes, sprayed the glass with venom whenever you walked by its holding tank. There were puff adders, Gaboon vipers, and mambas from Africa, and from Australia the small but nasty death adder. The worst of them all was the Australian brown snake, which was as fast as a red racer, extremely venomous, and, as if that wasn't enough, had a really bad attitude. These were all his "babies," as he called them.

There were giant snakes too. Although nonvenomous, they could be dangerous because of their sheer size and strength. Our sixteen-foot-long African python bit the herpetologist once when eight of us were moving it from one cage to another. We each had hold of a couple of feet of snake, and biologist Glenn Burghart had a firm grip on the head. The herpetologist walked around in front of the snake and it suddenly lunged for him, pulling Glenn with it. The python grabbed the herpetologist by the shoulder and wouldn't let go until it was pried off. This particular species of python has several long teeth meant for grasping, and one tooth must have hit a small blood vessel in his shoulder because blood was everywhere. Fortunately for me—I guess—I was at the opposite end of the procession, though at that end snake shit was everywhere, including all over me. The herpetologist never forgave Glenn for allowing the snake to bite him. The rest of us, however, were with the snake, secretly thinking a bit of justice had been done.

The herpetologist took great pleasure in feeding his charges. He preferred to use live food, whether necessary or not. A whole, live rooster went to each of the big pythons and anacondas. Rats, mice, "pinks" (newborn mice), and day-old cheeping chicks were fed to all the smaller species. He enjoyed throwing live white rats into the alligator swamp to see if any could swim to the side before the alligators got them. The herpetologist's feeding frenzies were definitely not something I liked to watch.

Because Herald wanted to squeeze as many exhibits as possible into our limited space, the venomous reptile exhibits were poorly designed from the standpoint of safety, with cages stacked in two tiers one foot apart. Cleaning the glass or removing snake shit required great skill and care to avoid being bitten. In order to work on the upper level of cages, you had to stand on a one-foot-wide plank two feet off the floor, which made jumping backward quite dangerous. Besides, the concrete wall opposite the cages was only four feet away, which meant there simply wasn't enough room to escape if an aggressive, venomous snake got out.

Periodically we had a snake-bite drill. When the alarm went off, we would time how many minutes it took to get from the aquarium to the University of California–San Francisco Medical Center, where the antivenins were kept.

Herald felt that in my position as assistant curator I should also be able to work with all the animals at the aquarium, and he asked me to begin working with the reptiles. I, however, did not agree. When Herald hired me from Marineland of the Pacific, he did so because of the marine exhibits I'd set up there. I was stubborn: if he hired me to do the aquatic exhibits, that is what I should do.

He told me to work with the reptiles, and I refused. Then he ordered me to do so. It wasn't that I was afraid of the snakes; I respected them, of course, but it wasn't what I had been hired to do. It was the principle of the thing. I went to Dr. George Lindsay, director of the Academy, and we agreed on a compromise: I would work only with the aquatic amphibians.

PROVOKING A SHARK ATTACK

Herald's attempt to have everyone do everything regardless of interest or ability eventually backfired, with near tragic results. At that time, aquarists Bob Kiwala and Lloyd Gomez, Phyllis Ensrud, who was Herald's secretary, and I were the only divers at the aquarium. Herald's next move, therefore, was to make every aquarist become a scuba diver.

Senior aquarist Tom Green was a good aquarist but was getting along in years and not exactly in top physical shape. Somehow, although Tom wasn't a strong swimmer, he passed the diving tests and was certified as a scuba diver. As far as Herald was concerned, he was now fit to perform any diving task.

At one point during the reconstruction, the large temperate marine fishes such as yellowtail, bass, and leopard sharks were taken out of their old exhibit tanks and placed temporarily in the newly completed fifty-five-thousand-gallon dolphin tank. Around this time, too, a four-foot-long sevengill shark (*Notorhynchus cepedianus*) was donated to the aquarium, and it was placed in the dolphin tank as well. Herald, who had taken a special interest in the new sevengill, insisted that we make sure it was fed daily. The tank was outdoors in the sun, and a resulting algae bloom made the water a pea green color with only three feet of visibility. This made it impossible to see through the window if the shark was feeding on any of the food tossed in for the other inhabitants. Herald therefore wanted us to dive into the tank, catch the shark, and bring it to the side for force-feeding. We would take turns with the diving, shark catching, and force-feeding.

I was standing ready with the food on one of Tom Green's days to catch the shark when suddenly he broke the surface with his arm held up in the air and blood pouring out. He made it to the side, and I saw a massive wound on the fleshy part of his forearm. I dropped his weights and tank and helped him out, sat him in a nearby chair, wrapped his arm in a towel, and ran for help. It was obviously a shark bite. An ambulance arrived quickly and took Tom off to the hospital.

As I hosed the blood off the wall and deck I glanced over at the tank and noticed something with hair on it floating on the surface. Grabbing a nearby dip net I fished it out; it was a piece of Tom's arm. After rinsing off the filamentous algae I raced downstairs, put it in a petri dish, and gave it to someone to drive to the hospital. The doctors sewed it back onto Tom's arm, but they didn't know until his cast was off six weeks later whether the attached piece would successfully graft. Luckily it did, and Tom regained almost full use of a badly scarred arm.

Tom told me later that when he finally saw the sevengill swim by through the murky water he grabbed its tail and it whipped around at him with its mouth wide open. He threw his arm up in front of his face, and the shark grabbed it and bit down hard. Even a small sevengill has a very large mouth full of small but razor-sharp teeth well designed for removing chunks of flesh. Tom is very lucky the shark didn't get his face. He never went in the water again, and Herald backed off on insisting that everyone be able to do everything.

The ongoing contest of wills between Dr. Herald and me, as well as an exciting new job offer, eventually led to my leaving after only three years at Steinhart Aquarium. By then Bob Kiwala had also left, having moved to Texas A & M University to work as a diving biologist on marine research projects in the Gulf of Mexico. Bob and I were destined to cross paths again later, but under much more favorable circumstances.

7

SEA WORLD AND SOUTH

The construction of a new oceanarium in San Diego, to be called Sea World, was the dream of four individuals, each with his own unique strengths. George Millay was the president and enthusiastic driving force of the project; the others were Milt Shedd, an investment broker and fishing enthusiast; Dave DeMotte, a creative accountant; and Ken Norris, the former curator at Marineland of the Pacific. Together they managed to raise enough capital to build and open the new oceanarium in 1964.

Sea World had been open to the public for a year when George Millay flew to San Francisco to visit Steinhart. He must have been impressed, because a few days later he called and asked if he could fly up and take me out for breakfast. Getting right to the point, he asked if I would come to Sea World as their curator of fishes. Although the present curator had been on the job since the beginning and was technically competent, and the systems and water quality were good, the exhibits themselves lacked imagination. George knew little about marine life, but he knew enough to know that the fish and invertebrate exhibits could be better than they were, and he wanted Sea World to have the best.

I'd heard disturbing stories about the hiring and firing of more than one of Sea World's general managers. I worried about leaving a financially secure, if frustrating, position at Steinhart and once again uprooting my wife and our young daughters to take a job that might be much less secure. Jobs for marine biologists who wanted to work in aquariums were few and far between. If this didn't work out and I ended up with no job, our family would be in a real fix. So I asked Millay for a two-year contract. He agreed, but I needn't have worried. I ended up staying at Sea World for nine rewarding years.

George Millay wasn't afraid to try something new, and I admired him for that. We had a clear goal: to strive for quality and uniqueness and to outdo our competition—Marineland of the Pacific, up the coast in Los Angeles. (When George was forced out by an organizational takeover a number of years later, Sea World lost much of the drive of those early years.)

Meanwhile, the new park struggled to survive. There were many days when the visitors coming in the front gate were anxiously counted to see if the week's payroll would be met.

Kent Burgess, the new director of training, had worked with Ken Norris at Marineland training the marine mammals. A graduate of the Keller Breland Institute of Animal Psychology, where he had trained birds, he used techniques based on B. F. Skinner's operant conditioning, which stresses positive reinforcement. Kent trained the new mammal trainers, and together they were able to develop some amazing behaviors with the dolphins, sea lions, elephant seals, and Shamu, the first killer whale in an oceanarium. I found the animal shows trite and anthropomorphic, but what the animals themselves could do was most impressive. Animal behaviors that today are accepted as no big deal were truly amazing then.

Eventually, through advertising and word of mouth, news of the new oceanarium in San Diego spread throughout L.A. and beyond and visitors came in increasing numbers. The place became a huge success.

BAJA CALIFORNIA

Apart from improving the aquarium exhibits, my first major responsibility at Sea World was to organize a collecting trip in 1966 to Cabo

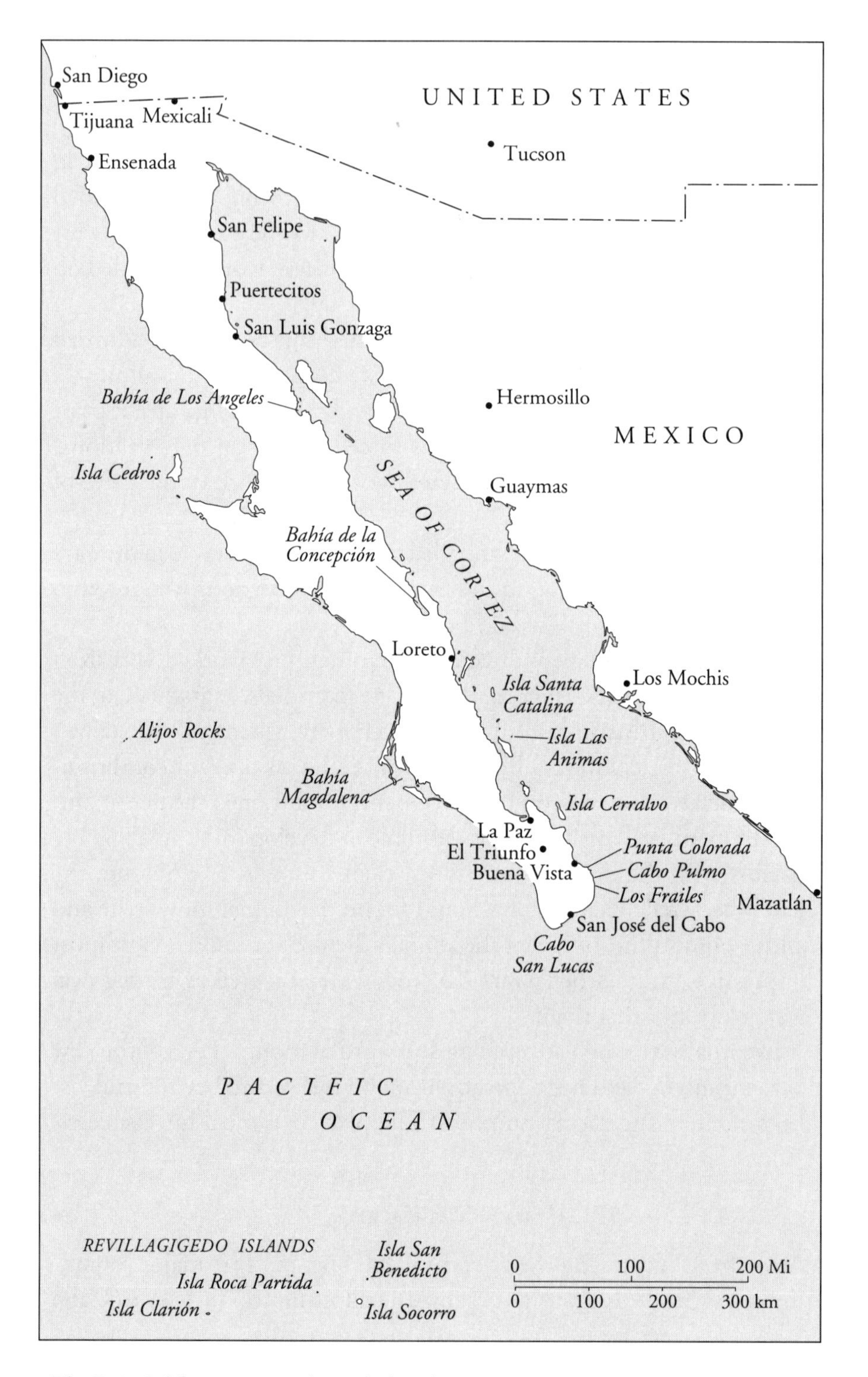

The Baja California peninsula, including locations mentioned in chapters 2, 4, 6–9, 11, and 19.

Cabo San Lucas, the southernmost tip of Baja California. (Photo courtesy Chuck Nicklin)

San Lucas, at the tip of Baja California. I'd been there once briefly on a side trip following the 1964 California Academy of Sciences Sea of Cortez expedition and had made a couple of solo dives in the precipitous Cabo San Lucas submarine canyon. At ninety feet I saw the famous "Rivers of Sand" falls that have scoured out the canyon. Evidently, they must pour over into the canyon only when triggered by earthquakes or storm waves because when I was there the motion was more a trickle than a river. High up on each side of the falls were the dead, scoured skeletons of sea fans, mute evidence of a massive sand avalanche in the not too distant past.

One memorable experience of that side trip—in addition to diving along the spectacular vertical walls of the canyon—was the oil pneumonia I contracted from breathing air pumped by an overheated scuba air compressor. It was June and the air temperature was 105; the poor little compressor simply wasn't getting the cooling necessary to condense the oil vapors. I had an oily taste in my mouth for six months afterward. I don't recommend it.

Among the friends of the founders of Sea World was a man named Willard Bell, of the Packard Bell electronics family. The family owned a 110-foot diesel yacht, *Five Bells,* which they offered to Sea World for the collecting trip to the Cape. Willard really wanted to help and at our request had modified the boat such that the seawater supplied to the holding tanks below deck could be heated with water from the engines.

On a collecting trip the previous year, Sea World staff had lost many of the more delicate tropical fish when the boat hit cold upwelling water off the west coast of Baja, just south of Ensenada. Willard's modification was intended to reduce the risk of this happening again. The control system was quite crude, however: it consisted of periodically sticking a thermometer in the holding-tank water and then adjusting a valve by hand—and hoping it was right. On the earlier trip, the fish merely ran the risk of dying from the cold. Now we had added the possibility of cooking them.

I'd been regaled with humorous tales from the previous year's trip. The best was a description of Milt Shedd, chairman of the board and an avid and skilled angler, sitting on the bottom in scuba gear trying to catch moorish idols (*Zanclus canescens*) with a tiny rod, reel, and minute baited hook. Moorish idols, of course, graze on the tiny animals and algae that encrust the rocks. While he didn't actually catch any, he certainly won the prize for effort.

We had assembled a wide variety of collecting gear, most of which I had never used before. We made wire fish traps and bought gill nets; in addition, we had every conceivable type of hook, line, and sinker (Milt was "the compleat angler") as well as fish anesthetic, dive lights, and, of course, my trusty underwater hand nets. With all that gear, our enthusiasm, and some skill, we were bound to get our quarry.

Before the trip I happened to talk to Susumu Kato, a U.S. Fish and Wildlife biologist who had spent considerable time in the Gulf and who had written the *Field Guide to the Sharks of the Eastern Pacific.* I casually mentioned that I was planning on collecting fish by night diving. He told me he'd caught twenty or so sharks one night off the end of the tuna cannery pier at the Cape. Around the pier at the Cape was right where we planned to work, and since I had never dived with sharks, especially at night, his news was a bit disconcerting. I really think he enjoyed telling me that and seeing my reaction.

Gear, food, and people were loaded on board and we had an uneventful three-day trip to the Cape—downhill, as they say, with the swell coming from Alaska. Finally we pulled into the protected lee of the spectacular tip of the long Baja California peninsula and anchored close to Sheppard Rock, next to the submarine canyon. We made our first night dive that same day after supper. As a way to find out what fish were there, we had first stretched a small gill net between Sheppard Rock and the shore, a distance of about one hundred feet. We climbed back on board and waited a half hour before going back down to check the net. What we found taught us instantly that gill nets are useless for collecting most aquarium fish.

Among the various fishes caught in the net there must have been twenty beautiful Mexican goatfish (*Mulloidichthys dentatus*), which, in their struggles to get free, were shedding their large scales like confetti. There were also a couple of porcupinefish (*Diodon holocanthus*), an animal that, when threatened, gulps water and inflates itself to the maximum. Of course, this defense strategy was quite useless in the present circumstances; in fact, the fish were now virtually impossible to remove without severe damage to them, us, or the net from their needle-sharp spines. More interesting than what was in the net, however, was evidence of what had gone right through it: a giant hole, large enough for a VW Bug to drive through. All of a sudden we had visions of Sus Kato's nocturnal sharks. We never did know what made that hole, but whatever it was, it was really big, strong, and probably not something we wanted to meet underwater at night!

Sus Kato's shark story and the presence of the *tiburoneros,* or itinerant shark fishermen's camps, were concrete evidence that sharks were definitely around. Yet, for whatever reason, I haven't seen a single shark during the hundreds of night dives I've made over the twenty-five years of collecting around Baja California. Other trips, 250 miles south at the island of Socorro, in the offshore island group the Revillagigedos, were to be a different story.

We learned a lot on this first collecting trip to the Cape, and it was truly a luxury to have the *Five Bells* as our base of operations. A key member of the team was Leo Navarro, a retired tuna fisherman who, among his many skills, was an excellent cook and prepared delicious Italian meals for us. He was also an expert net mender and many a time repaired our damaged hand nets.

We made our share of mistakes, but we learned from them and developed some successful methods for catching the elusive reef fishes, including Milt Shedd's coveted moorish idols. Drawing on my past experience collecting for Marineland of the Pacific, and especially observations of fish behavior I'd made while diving for lobsters, I chose the combination of night diving and hand netting as the most effective way to collect diurnal, or day-active, fishes. Butterflyfishes, angelfishes, parrotfishes, and moorish idols are all much too active and wary to approach during the day. At night they become semi-dormant and are easily confused—and collected—when a bright dive light is shined in their eyes.

We each carried a small mesh-nylon hand net, a plastic bag under the crotch strap of our wet suit, a dive light, and a number of different-sized hypodermic needles tucked under the wrist cuff of our wet suit. The hand nets were our most prized possessions; they couldn't be purchased, so we each made our own, with our own details of design and construction that we swore were better than anyone else's. (Of course, my design really *was* the best of all!) The hypodermic needles were to release the expanding air from the swimbladder of fish brought up from water deeper than thirty or forty feet, relieving buoyancy and stress.

As more and more night dives were made we developed skills and techniques that increased our effectiveness. We learned to hold our breath when approaching a fish so as not to startle it with exhaled bubbles. Fish may not be able to see in the dark, but they can certainly hear and sense sounds and vibrations quite well.

Our buddy system was simple, although it would be deplored by current occupational safety standards. The absolute rule was this: we were together in the boat before the dive and we were together in the boat after the dive. While in the water, though, we each pretty much went our own way. Sticking together in pairs would have cut our catch rate considerably, especially in the dark. We would have wasted too much time looking for our buddy in the dark instead of concentrating on catching the fish. And catching fish was our primary concern.

The easy way to start a dive: Bob Kiwala, his collecting net and bag in hand, just leans back and lets gravity do the rest. (Photo by author)

A little competitiveness to see who came up with the most helped motivate us to roll over the side of the boat into the pitch-black water.

Hand-netting fish at night is an art that develops with practice, and we each came up with our own techniques. If you shine the dive light right into the eye of the chosen fish, it can usually be netted quickly with a flip of the wrist. Then the job is to get it out of the net and into the plastic bag tucked under your crotch strap. The dive light is usually held between your legs while the bag is pulled out from your suit. Then, grasping the fish in the netting, you invert it into the bag and shake it loose, without letting any already collected fish escape. This procedure is accomplished without being able to see much or even to swim because the light between your knees keeps you from kicking. If your buoyancy is a little negative, there's always the risk of sinking down on an unseen patch of needle-sharp sea urchins. Certain fish, like the barberfish, would freeze when hit with the light and on very dark nights could actually be pinned against the front of the light with your hand, no net needed.

Although night collecting was effective, we learned that it worked well only if it was really dark. The darker the night, the better. A little bit of moonlight was just enough for the fish to see and elude you, and collecting during a full moon was pretty much hopeless. Because darkness was the key to our success, we'd time our dives before the moon came up or after it went down. That sometimes meant we'd go out for our dive at three in the morning and finish just at first light. In one respect this was good, since then we were putting our catch of the night into the floating fish receivers in the light of dawn rather than in the dark.

Another reason to go diving in the wee hours of the morning, we found, was to avoid the major heartburn that Leo Navarro's delicious but rich Italian cooking gave us when we went diving right after supper. Underwater burps that tasted like lasagna, garlic bread, and red wine were more than we could take.

THE PARROTFISH CHALLENGE

There are four species of beautiful parrotfishes in the Sea of Cortez. The mature male of the bumphead parrotfish (*Scarus perrico*) has an impressive hump on his forehead and a fuzzy dark growth on his upper, parrotlike beak that resembles a mustache. We were of course eager to catch these striking fish, but at a length of two and a half feet and weight of twenty pounds, they were quite a challenge to collect.

During the day, actively feeding parrotfish are alert and virtually unapproachable. There's no sneaking up on them, so trying to hand-net them was out of the question. Almost as if they could read your mind, they'd take one look at you and be gone in a flash of green. Being coral and algae grazers, they couldn't be collected with hook and line or traps because they wouldn't take a baited hook or go into a baited trap. Gill nets were also out of the question because of the damage the nets do to the fish's skin and its protective mucus layer.

At night parrotfish find a hole or an overhanging rock ledge and bed down until morning. Many secrete a loose mucus cocoon that envelops their entire body. The function of this cocoon is not fully understood, but one theory is that it somehow protects them from potential predators like sharks that locate their prey by their sense of smell. At night the resting parrotfish can be easily approached and sometimes even gently touched. They looked like easy targets for our hand nets. We tried but had grossly underestimated the speed and power of a rudely awakened parrotfish.

Kelly McColloch was the first to give it a try. He placed his net in front of a big sleeping parrotfish's head and gave it a quick goose in the tail. The parrotfish took off like a rocket and was gone, leaving a gaping hole in the end of Kelly's precious net. The strong nylon didn't even slow it down.

Never ones to be beaten by a mere fish, we put our supposedly superior intelligence to work designing a special parrotfish-proof hand net made out of heavy-duty braided nylon mesh of the type used to capture dolphins and killer whales. The conspicuousness of the heavy netting wasn't a problem because it would only be used when it was dark. This time the net held, though for a few moments before we could get it into our net collecting bag we found ourselves being dragged around by the furiously swimming fish.

Swimming back to the boat in the dark carrying a large, angry parrotfish was one of the few occasions I thought of sharks in the Gulf while night diving, the ability of sharks to pick up on the vibrations put out by struggling fish being well known. All's well that ends well, however, because we never saw a shark. I'm sure there must have been sharks that saw us, but apparently we didn't interest them.

After a few days of collecting, we had another chore to attend to: the fish we had already caught now needed to be fed. This became a daily job and involved much hook-and-line fishing for other species to fillet, chop, or grind up and toss into our floating aquariums. By the time the receivers had a few hundred fish, our days and nights were pretty full.

Although night diving and hand netting was an effective collecting method, it wasn't practical for all fish. Some species could be collected only during the day. The methods we used to catch the diurnally active fishes worked in reverse for the large-eyed nocturnally active fishes like squirrelfish and cardinalfish. They're the night shift on the reef. They take over when the sun goes down and are out actively foraging for food after dark. As soon as dawn comes, they retreat to a dark cave or crevice and wait quietly for nighttime to come again.

Taking advantage of this behavior, we would squirt a dilute solution of the fish anesthetic quinaldine into their daytime hiding place. They're reluctant to come out of their hiding place into the bright daylight, and if it was done slowly enough they'd be knocked out before they realized what was happening. The result was sometimes a veritable rain of upside-down, zonked-out, immobilized red squirrelfish (*Myripristis occidentalis*), soldierfish (*Adioryx suborbitalis*), and cardinals (*Apogon retrosella*). The trick then was to gather them quickly with a gloved hand into the plastic bags before either they woke up or the morays grabbed them for their lunch.

Soldierfish are much too spiny to use hand nets or bare hands on, something we found out early on after spending five minutes untangling one soldierfish from our hand net while all the others recovered and swam away. An occasional unconscious soapfish (*Rypticus* spp.) might drift out of a crevice with the others, but we learned the hard way not to put a soapfish in a plastic bag with other fish. When irritated—and we certainly did irritate them—soapfish can secrete a foamy, toxic mucus that makes them unpalatable for a predator to eat. In the close confines of a plastic bag, this mucus can kill everything in with them. They are fine in a large aquarium with other fishes, just not when trapped in a plastic bag.

All members of the wrasse family of fishes, like the beautiful rainbow wrasse (*Thalassoma lucasanum*), which is named for Cabo San Lucas, and the green and orange sunset wrasse (*T. lutescens*), go to sleep at night either beneath the sand or hidden in the coral or cracks in the rocks. We developed a simple method for catching the rainbow and sunset wrasses. It involved empty one-gallon, clear-glass mayonnaise and pickle jars that we had collected before the trip from dumpsters behind restaurants, and a few black sea urchins to use as bait. In addition, it relied on the fact that wrasses, which are very visual animals, will not enter a space where they don't have a clear view of what's happening all around them.

Here is how it worked. In the shallow water that the wrasses inhabit, near coral-encrusted rock reefs, we broke open a sea urchin, and the feast began. Rainbow wrasses appeared out of nowhere and, fearlessly, descended by the dozens on the free meal. We then broke open another urchin and popped it inside the glass jar. Soon one fish summoned up enough courage to venture inside for the tempting snack. The others saw it eating and they rushed in as well. At that point we slammed the lid over the jar mouth, and we had our wrasses. The jar of frantic fish was then taken ten or fifteen feet up to the boat overhead and poured into the waiting plastic trash cans we use for holding fish. This was repeated until we had a couple of hundred wrasses.

We held the wrasses in sand-bottomed tanks on the mother boat. At night they'd disappear from view with just the tips of their mouths protruding from the substrate so they could breathe. They don't do well if they have to sleep in a tank with a hard, bare bottom or in a rough net receiver. Other aquarists have built clear-acrylic traps with hinged doors and tapered-entrance funnels, which are elaborate and expensive works-of-art, but I think plain old pickle jars plus the ability to read the fish's behavior outproduce the complicated acrylic traps. Unfortunately, modern petroleum-based technology has made it hard on us low-tech collectors: it's impossible to find one-gallon glass jars anymore; everything now comes in opaque plastic, which is no good for catching fish.

YELLOWTAIL SURGEON ROUNDUP

One of the most striking sights in southern Baja California is a school of large yellowtail surgeonfish (*Prionurus punctatus*) cruising over a rocky reef. Through trial—and some error—we developed an interesting method of capturing these fish.

Locating an area where they habitually cruise, we stretched a one-hundred-foot-long, ten-foot-high barrier net perpendicular out from the rocky shore. Fairly soon, a school of twenty to fifty surgeons would come moseying along the reef, grazing on algae. Pretending they were the last thing on our minds, we'd casually swim around behind them, and then Kelly McColloch, Bob Kiwala, and I would slowly herd them like grazing sheep toward the net. As soon as they saw the barrier they'd turn to double back, but we were blocking their path. Then things would get wild. As the three of us rushed at them, frantically waving our outstretched arms, the fish would panic. The smart ones managed to shoot

A school of yellowtail surgeonfish cruising the reef. (Photo courtesy Robert S. Kiwala)

right between us and get away, but many of them would lose their heads and charge straight into the net.

Yellowtail surgeons have three sharp blades protruding from the base of their tail—hence the family's name. Not as scalpel-sharp as on the single-blade species of surgeonfish, these blades would tangle in the net long enough for us to grab them with our hands, untangle them, and transfer them to our collecting bags. Unlike with most fish, the skin of yellowtail surgeons is tough and leathery, so they are one of the few fish that can be netted like this without any damage to the skin, fins, or protective mucus layer.

Although it was exhausting, we enjoyed the thrill of the chase when we collected yellowtail surgeons. On a good set we could catch twenty or thirty nice big specimens. Their one-and-a-half- to two-foot-long bodies are gray with many small black spots, and of course, they have yellow tails. Just as in the ocean, they make a striking school in a large aquarium.

A garden of eels. (Photo © Norbert Wu)

A GARDEN OF EELS

Among the most extraordinary fishes in the Sea of Cortez are the garden eels (*Taenioconger digueti*), which live in colonies at moderate depths of thirty to eighty feet on a sandy bottom. As the name implies, they literally form a garden of eels.

Growing to about two feet as adults, they live in burrows. When feeding and undisturbed, they extend half or more of their body length out of their burrow and gently sway and weave as they search with their sharp little eyes for tiny planktonic animals drifting toward them in the current. Some areas, such as the west side of Isla Cerralvo, southeast of La Paz, have what must be millions of garden eels in colonies half a mile long. As you swim toward them, they imperceptibly glide down into their sand burrows and disappear. Then when you look behind you, there they are slowly coming back up again. It's almost magical. They're found throughout the southern Gulf of California, and there are other species in other parts of the tropics.

Kelly McColloch with our first successful garden eel collection.
(Photo courtesy Robert S. Kiwala)

Garden eels are one of the most challenging animals to collect. Early attempts to do so by squirting quinaldine anesthetic down their burrows failed. It may have put them to sleep, but it was practically impossible to dig them out of the loose sand. Eels also have a remarkable ability to swim backward through sand.

We eventually came up with a successful method. It consisted of two divers going down to the eel colony with three or four sheets of eight-foot-square clear-polyethylene plastic and two squeeze bottles of quinaldine. A plastic sheet was carefully spread over an area that had a large number of eel holes, with the edges tucked under the sand to keep the sheet down on the bottom. The end result was a low plastic tent, into which some anesthetic would be squirted. The divers would then move on and lay out the rest of the plastic sheets in a similar manner. This task pretty much consumed the two divers' air supply, and they would return to the boat. A second team of two divers would then take over and descend to the eel tents. By now some of the eels were partly out of their holes but not yet knocked out enough to remove. The divers would squirt a little more quinaldine under each of the plastic sheets and wait a few more minutes. Returning to the first sheet, they could now collect the very groggy eels lying prone on the sand. They would simply lift off the tent and gently pull the eels out of their burrows, placing them in plastic bags.

Because they're sand dwellers, like the wrasses the garden eels must be held in tanks with a deep bed of sand for them to burrow in. They're a challenge to collect but make a fascinating aquarium exhibit. Little is known about their biology. How do they reproduce? Why do they live in such large colonies, and how do they get there in the first place? Now that a reliable method of collecting them has been developed, answers to some of these questions may be learned in an aquarium.

OUTWITTING A CAGEY FISH

Another interesting fish we sometimes saw swimming over shallow sandy areas was a species of wrasse called the razorfish (*Xyrichthys pavo*). It gets its common name from the sharp edge to the front of its head. We discovered the function of this sharpened head when we tried to catch this fish with hand nets. Before we got within netting range, the fish would dive into the sand and disappear. Digging right where it had gone in was fruitless: the fish simply wasn't there. The sharp wedge-shaped front of its head enables the fish literally to swim through the sand and pop up some distance away.

Collecting the slippery razorfish by diving obviously wasn't going to work, so Leo Navarro, a nondiver but an excellent fisherman, took on the challenge of capturing this elusive animal. He took the skiff, a hand line of thin monofilament, some tiny hooks, and some fresh shrimp for bait. He then spent hours peering over the side into the clear water, maneuvering the baited hook near razorfish he spotted cruising over the sandy bottom. His patience won out and he collected a number of these attractive light-blue wrasses. We set them up in a tank with sand so that, like other wrasses, they'd have a place to sleep at night.

The elusive razorfish in their new aquarium home. (Photo by author)

On the return trip to Sea World we found out the hard way what their favorite food is. Because holding space for all the fish we were bringing back from Mexico was tight, we decided to combine two of the species of sand-dwelling fish—the garden eels and razorfish—in one tank just for the night. The next morning we found each razorfish with part of a garden eel sticking out of its mouth like a cigar. It was no problem for them to catch a garden eel—all they had to do was swim into the sand. As mad as we were about losing some of our precious, hard-won garden eels, we couldn't help but laugh at the sight.

UNHARMED AND HEALTHY

Although there are hundreds of books about how to catch fish, beginning with Izaak Walton's *Compleat Angler,* practically nothing has been written about how to collect fishes and invertebrates alive, unharmed, and healthy. You either figure it out yourself or you go with someone who in turn learned it from someone else who figured it out on their own. Seeing a new creature, the collector's mind immediately begins to figure out a way to collect it.

Some techniques are developed by chance rather than by plan—as in the case of the beautiful but aggressively territorial blue-spotted jawfish (*Opisthognathus rosenblatti*), found north of the Cape in the Gulf of California. These four-inch fish, like all species of jawfish, are named for the large mouth they use to brood their eggs and to dig and construct their elaborate burrows in the gravelly bottom. Excellent structural engineers, they reinforce the sides of their burrow with interlocking pieces of shell or rock so the burrow doesn't cave in on them. Materials of just the right size and shape are highly prized, and they're constantly stealing choice pieces of building material from each other.

One method we found effective with jawfish takes advantage of their industrious nature and their habit of constantly working on their burrow. A short piece of monofilament fishing line is tied to a small, unbaited hook. The hook is then dropped down their burrow. Not about to tolerate a strange object in its home, the fish picks up the hook in its mouth to remove it. A quick jerk sets the hook and the fish is pulled out and popped into a plastic bag. The small hook wound heals quickly, and when placed in an aquarium the fish quickly sets about constructing a new burrow.

The Gulf of California is home to the world's largest species of jawfish, appropriately named the giant jawfish (*O. rhomaleus*), which reaches a size of twenty inches. I learned just how good these fish were at construction when I attempted to start a burrow for three jawfish we brought back. With a good assortment of shell fragments and small rocks I tried to reinforce the sides of a burrow I dug in the gravel in their new tank. It kept caving in and I finally gave up in disgust.

The fish, however, soon got to work, picking up pieces of shell and rock with their mouths to build their own burrows. It wasn't long before their burrows were complete, with solidly reinforced walls. Most of the action occurred out of our sight, so we don't know how they do it, but they certainly deserve an advanced degree in engineering.

Some fishes, we found, were drawn at night to the pool of light surrounding our safely anchored boat. First, assorted planktonic animals swam toward the light; these in turn attracted small fishes, then larger

The first scythe butterflyfish ever to be brought up live from the depths of the Cabo San Lucas canyon. (Photo by author)

fishes. This gave us an excellent opportunity to collect the iridescent silvery lookdowns (*Selene brevoorti*) and moonfish (*Vomer declivifrons*). A leader rigged with five very small hooks with attached colored yarn called "Lucky Joes" was effective at catching these beautiful fishes of the jack family.

The Cabo San Lucas submarine canyon is only fifty yards or so from the shore, and we made a number of deep dives below the sand falls and into the canyon. This was one place we did use the buddy system. The canyon wall becomes almost vertical at a hundred and fifty feet; at that depth, under the euphoric effects of nitrogen narcosis, you feel great and the temptation is strong to keep going ever deeper. It was on one of these dives, at the mind-numbing depth of one hundred and eighty-five feet, that Bob Kiwala captured a live deepwater scythe butterflyfish (*Chaetodon falcifer*)—a first-time feat. At the surface, though with a few needle holes from Bob bleeding the air out of it on the long way up, it looked quite healthy. We were ecstatic, and a few days later

A fine collection of Mexican tropical fish on board the *Five Bells.* (Photo by author)

drove all the way to the La Paz airport to fly Bob's precious butterflyfish alive back to San Diego and the Scripps Aquarium.

On our eventual return we learned to our dismay that Scripps ichthyologist Dick Rosenblatt had promptly pickled it in formalin the minute he got it! As a taxonomist, Dick thought the fish had much more value as a scientific specimen in a collection jar than as a fish swimming in an aquarium. Of course, we, as aquarists and lovers of live fish, didn't see it the same way at all. Anyway, that was the end of the first scythe butterflyfish to be captured alive. Both Bob and I muttered into our beer about that one for months afterward.

SAFELY HOME

The first Cabo San Lucas collecting phase ended with the all-day task of carefully transferring the hundreds of fishes from the floating receivers to the below-deck holding tanks on the *Five Bells.* Next we had to break down the receivers and secure them on deck. That done, the boat was fueled up and made ready for the uphill trip against the northwest swells back to San Diego.

This was the part I dreaded most about these expeditions. Not only were we dead tired from the hard work of loading the boat, but then we had rough seas and queasy stomachs to look forward to while standing wheel watch at all hours. What's more, no matter how rough it got or how bad we felt, our precious cargo of fish needed frequent, around-the-clock monitoring of water flow and temperature.

Without fail, we would hit heavy seas just north of Isla Cedros. On one notable voyage, after the railing had been torn off, the skipper decided that it was getting too rough for the boat and we should turn around to seek shelter behind Cedros and wait for the seas to quiet down. The *Five Bells,* though 110 feet long, is quite narrow. Turning broadside to those swells and heavy with full fish and fuel tanks, we ran the very real risk of rolling over. Fortunately we made it, but everything not securely bolted down went flying, and two of the marine radios came off the wall and smashed on the floor, leaving us without communication. But at least we were safe behind the island, with a chance to secure everything that had come loose from the recent pounding.

In two days the seas had calmed down enough for us to head north for San Diego and Sea World. We walked, or rather staggered, off the boat, leaving the formidable job of unloading the fish to the more rested aquarists and other Sea World staff who pitched in to help. The majority of our precious cargo was destined for the fifty-thousand-gallon Sea of Cortez tropical exhibit, while the smaller species would be on exhibit in the tanks of the marine aquarium.

The next day everyone in the park came to see the results of the expedition. It was truly spectacular! There were huge, bright green bumphead parrotfish, hundreds of barberfish, regal-looking Cortez and king angelfishes, and a host of other colorful fishes. This was a sight never before seen in an aquarium—a big exhibit full of the wonderful fishes of Mexico's unique Sea of Cortez.

8

CARNIVAL IN MAZATLÁN

THE SEA OF CORTEZ EXHIBIT was very popular with Sea World's visitors. As with all exhibits that feature live animals, we occasionally had to replenish the collection, but fortunately not in the numbers of that first trip. By 1967, the year we needed to make another collecting trip, Willard Bell's generosity had put a significant dent in his bank account and he couldn't afford to be as generous a second time. He did volunteer to run his boat down to the Cape and transport fish back to Sea World, but he couldn't afford to pay his crew to stay there for the four weeks we would be collecting.

A new and much lower-budget plan was hatched. We'd drive to the Cape with our collecting gear piled onto a pickup truck. At that time, of course, the only road down the peninsula of Baja California was an extremely rugged one-lane dirt track. Even if we could make it to the Cape via the Baja "road," it was questionable if the truck, our equipment, or we would still be functioning once we arrived. The new plan was to drive to the Cape via the paved highway on the mainland of Mexico. We would cross the border into Mexico at Mexicali, then head south to Guaymas and on down to Mazatlán. At Mazatlán we'd take the big ferry across the Gulf to La Paz and drive the remaining 120 miles to the Cape.

The team this time was Sea World aquarist Kym Murphy (now with Walt Disney's Imagineering); Bob Kiwala, the collector at Scripps; and me. We took off, truck piled high with scuba gear, an air compressor, folded-up fish receivers, pesos for the trip, and enough Budweiser to get us to the border, where we could get some good Mexican beer. We had everything under control, but there was still one unknown. In those days it wasn't possible to make a reservation in the States for the truck to get on the ferry; you could only do so at the ferry office in Mazatlán. *No hay problema, Señores,* we were told.

It's 1,300 miles from San Diego to Mazatlán. On U.S. roads with three drivers, we figured, the trip would take less than two and a half days. In that part of Mexico, however, it is dangerous to drive after dark; the highway down the west coast of Mexico is not fenced, and cattle and horses are free to roam. Unfortunately for the animals, in this arid country where food is scarce at best, the choicest grass grows right along the edge of the pavement, watered by occasional runoff from the highway.

At sixty-five miles per hour at night, there's no way you can stop in time if a black cow decides just as you're coming over a hill that the grass is greener on the other side of the road. That would spell the end of the cow, the truck, and the collecting trip. The big Mexican trucks drive at night, but they protect their grilles and radiators with heavy pipe structures welded on the front. This doesn't do much for the poor animal, but it does save the trucks. A flock of vultures feeding by the side of the road is a frequent sight south of the border.

It struck me that natural selection must be in favor of the far more visible white cows and against black cows in Mexico. So much for selective breeding of desirable color varieties of cows and horses through careful husbandry; now they're being selected through differential death rates from the Mexican big rigs.

We stopped the first night in the city of Hermosillo and the second night in Los Mochis, where we made an observation-only visit to the red-light district to check out the nightlife. Getting up early on the third day, we gassed up the truck and headed south for Mazatlán.

Arriving on the outskirts of Mazatlán we noticed small groups of Mexicans, attired in white suits, walking along the roadside toward town. Many were carrying guitars. Thinking nothing of it, we headed for the ferry office. We had work to do, and the sooner we got across to La Paz, the sooner we could start collecting.

We were greeted at the ferry terminal by a smiling clerk who informed us that there was no space on the ferry for four days. Seeing our dejected looks, he told us, "Eet is no so bad—Carnival begins tonight!"

Thinking back, I don't really believe there was no room on that ferry. I have a strong suspicion they just tie the ferry up and join the party until Carnival is over. Come to think of it, that's not a bad attitude toward life. We could probably learn something from them.

In any case, we were stuck. What could we possibly do for four days in Mazatlán while we waited for the ferry? The answer was obvious: Forget about the delay and throw ourselves into it and enjoy! And we did—or at least Bob and I did. Poor Kym was miserably sick with the "turistas" and spent his entire Carnival days and nights alternating between moaning on his motel bed and running to the bathroom. However, his moaning failed to dampen our spirits, and we left him there and caught the bus into town to experience Carnival to the fullest. I hadn't budgeted for this kind of extracurricular activity, but in an honest attempt to save money we were able to find the largest, cheapest shots of tequila in all of Mazatlán. From what I remember of it, the music, singing, and dancing were truly wonderful. The Mexicans knew how to forget about their hard life, at least for a little while, and they really let loose and enjoyed themselves.

The highlight of the fiesta was the fireworks show on the last night. A tall, rickety bamboo structure had been assembled in the center of the main plaza, and as soon as darkness fell fireworks of all shapes and colors began whirling, sputtering, and shooting off in the most unexpected directions. For the grand finale the entire top of the bamboo tower, propelled by attached rockets, lifted off and disappeared over the roofs of the nearby buildings. There seemed to be little concern as to where or on whom it might land. The whole scene was wonderful.

With Carnival finally over, the next morning we made our bleary-eyed way to the ferry terminal and waited patiently in line with all the big trucks carrying supplies across the Gulf to La Paz and Baja California Sur. You learn to be patient in Mexico. Scheduled events seldom take place on time—a fact that will drive you crazy if you hang on to the American expectation that things happen when they're supposed to happen.

Our turn eventually came and we drove on board and down into the bowels of the big ship. The ship was crowded: people—families, truck drivers, businessmen, a few tourists, and the three of us—were sleeping anywhere on the upper deck they could find a place to lie down. Before the paved two-lane Baja highway was completed in 1972, the ferry was the only practical way to reach Baja California Sur.

Arriving in La Paz, we shopped for groceries and, of course, more cases of *cerveza,* then headed south to the Cape. Kym, still feeling awful, lay on the nets in the back of the truck and added external sunburn to his internal troubles. We stopped briefly at El Triunfo, a picturesque abandoned silver-mining town, and after that took a refreshing cold-beer break at the bar of the first sportfishing resort in Baja, Buena Vista, where the wealthy fly in to catch marlin. Then onward to the little house of a resident American diver, Don Scott, and his charming Mexican wife, Julia, where we had arranged to stay while at the Cape.

We quickly discovered that land-based collecting is a lot more work than collecting from a boat. The first job was to put the fish receivers together on land, drag them to the water, and then scrounge suitably large, heavy objects to use as anchors.

We had the use of an old thirteen-foot Boston Whaler, formerly a dolphin chase boat for a yellowfin tuna purse seiner, that was tied to a mooring a hundred feet or so off the beach near the tuna cannery. To keep the front end down at high speed in choppy water, the bow had been filled with concrete. It was a beat-up old scow, but it still floated—testimony to the abuse the rugged Boston Whaler can take.

We'd been used to having our dive boat tied alongside our mother ship, gassed up, ready for us to step in and go. This new arrangement was quite different, and a lot more work for just us three. None of us relished swimming out on a pitch-black sea at night to bring the Whaler in to the beach to load our dive gear, and we kept very close track of whose turn it was. Even though none of us had seen a single shark on all our previous dives at the Cape, we each had visions of sharks beneath us as we swam out through the black water. After our dive, too, we had to take the boat back out to its mooring and then swim back in to the beach. While diving at night never bothered us, there was something unnerving about swimming on top of the water in the dark. Each time it was my turn to deal with the boat, I felt quite vulnerable.

Our days were spent catching food for the fish we'd collected and repairing the net covers on the receivers to keep out the vigilant and ever-hungry pelicans. For hours we'd have to listen to the infernal racket of the gas-powered scuba compressor as it refilled the tanks we'd used the night before. The compressor was slow, noisy, and so heavy it could barely be called portable, but it was invaluable for our work.

At one point during this trip Betty hitched a ride down to the Cape and back in the private plane of a man coming down on business. She made a couple of scuba dives with me around Sheppard Rock, so she had a chance to see what these beautiful fish look like and where they live.

Collecting was good, and we had a large population of fishes in the receivers when the *Five Bells* came down from San Diego to pick up the fish and take them back to Sea World. Soon thereafter we loaded the pickup truck and headed north to La Paz, the ferry, and the long road home to California. Getting back into highway traffic was definitely a culture shock after the quiet, rugged beauty of the Cape and its rich surrounding desert.

MISADVENTURES IN LOS CABOS

We needed to make another trip the following year, 1968, to collect a few more of certain species. We had agreed among ourselves to move our operation to some location other than Cabo San Lucas to avoid the possibility of depleting the fish population, and had discussed the suitability of alternate sites. A good location needed to have road access, a place to stay that was at least one step up from sleeping on the beach, and an anchorage where the fish receivers were protected from ocean swells. We also needed to be where the big boat could come in later to pick up the fish.

We decided that a place called Los Frailes came closest to meeting our needs. Los Frailes is around the point from Cabo Pulmo, glowingly described in John Steinbeck and Ed Ricketts's wonderful philosophical and biological account of their 1940 expedition to the Gulf, *Sea of Cortez.* Cabo Pulmo is one of only two coral reefs in the Sea of Cortez; Los Frailes is the other one. Because of its uniqueness, in 1995 Cabo Pulmo was designated a marine park in which the taking of all marine life is prohibited.

There were no accommodations at Los Frailes, but thirteen dirt-road miles away was Punta Colorada, a small sportfishermen's resort run by Bob Van Wormer. Bob Kiwala had become friends with Van Wormer in 1964 when Bob towed a two-person submarine down from San Francisco to Punta Colorada; there he and Ted Hobson, a UCLA postdoctoral student, used the submarine to do night and day behavior observations of the reef fishes of the Gulf. Van Wormer would give us a reasonable rate to stay at his place, which was close enough to the collecting area for us to commute. Everything seemed to be working out fine, and we were all looking forward to collecting at Los Frailes. The team this time was Bob Kiwala, Kelly McColloch, and I, plus a friend of Kelly's, Ray Szymczak, who planned to drive down in his own Datsun pickup truck just for the fun of it.

We left San Diego, the truck piled high with the usual collecting gear, and headed south to Mazatlán and the ferry. This time, however, we had cleverly found out the dates of Carnival and, unbeknownst to Sea World or Scripps, had planned our San Diego departure so we would "just happen" to arrive in Mazatlán on the first day of the celebration. Sea World management never did find out about that little scheme.

Four days later, we somehow made it to the ferry and across the Gulf to La Paz. After stocking up at the *mercado* and making a stop at the cerveza distributor, we drove south toward Punta Colorada. A few miles south of the Buena Vista fishing resort it began to sprinkle. This was

Left to right: Author, Bob, and Kelly, our plans ruined by unexpected rain that turned the dirt roads into a morass of mud. (Photo courtesy Ray Szymczak)

an unexpected occurrence for this time of year, but we didn't think too much about it. Just a shower, we supposed. The rain got heavier, though, and pretty soon we were having to make a run at it to get through some of the flooded arroyos that crossed the dirt road.

We didn't worry until Ray's little Datsun got halfway through one, drowned out, and was stuck. A Mexican road crew was watching and thought it was the funniest thing they had seen all week. They finally took pity on us and hooked Ray's Datsun to their big road grader and pulled him out. We thanked them, shared some of our cerveza with them, and were on our way again.

Eventually we arrived in Punta Colorada in the rain, where we were greeted by the sight of the resort's sportfishing boats being torn loose from their moorings by big waves and washing ashore. Despite valiant, but futile, efforts to keep the boats from coming ashore, they hit the beach, breaking off outdrives, props, and rudders. It was a severe financial setback for the people of Punta Colorada.

The next morning we headed out to Los Frailes on a reconnaissance trip. The dirt road had now turned to slippery mud, and the thirteen miles took two hours. Our well-thought-out plan was falling apart; clearly, collecting at Los Frailes was impossible. The hard work of collecting and keeping everything together was enough without having to fight a barely passable road two or more times a day. What to do now? Here we were in Baja with no means of communication and no alternate place to work.

The pressure was on. We knew we had to get the fish, and we were not about to concede defeat and just go home. We hated to do it, but rather than pick a totally unknown place to set up base camp and collect, we decided to go back to the Cape, even though we had planned to leave the fish there alone.

The lack of any other protected cove to anchor the fish receivers was the main factor in our decision. Settling for an unprotected place would have been an invitation to disaster. The first ocean swells that came in, if they did not simply tear the receivers loose to wash ashore, would otherwise have exhausted and eventually killed the poor fish huddled inside the constantly moving netting.

Getting back on the main graded road, we found the rain had been quite local: toward the Cape everything was dry. A heavy storm swell, however, was crashing onto the beaches, and boats all along the south coast were being washed ashore and destroyed. We found out later that more than fifty sportfishing boats were lost in Baja from this unexpected storm.

We arrived at the Cape with no place to stay. The only hotel at Cabo San Lucas, the Hacienda, was much too expensive for our modest budget, and Don Scott, the former Mr. Arizona and macho bodybuilder, was no longer in the area. While his wife, Julia, was away visiting relatives he'd sold their belongings, taken the money, and skipped the country back to the States. Julia was fortunate enough to get a job as a maid at the Hacienda Hotel. That whole situation made me ashamed to be an American.

In the process of asking around for a place to stay we were told that we might be able to find accommodations at the "Clinique." The Clinique was owned by a Beverly Hills doctor who supposedly set it up to provide free medical treatment for the residents of Cabo San Lucas. He was rarely there, and when he did come he flew down in his own plane with his friends and partied. He was hated by the residents and the local Mexican doctor, who saw through his scam of getting a U.S. tax deduction by pretending to be a Good Samaritan. His caretaker gave me his Beverly Hills phone number, and I reached the doctor through the marine radio at the Hacienda Hotel. I managed to convince him to let us stay at the Clinique for the duration of our collecting trip.

The place had beds, a kitchen, its own gasoline-powered generator, and a caretaker—but a distinct lack of much in the way of functional medical equipment, which confirmed what we had been told. Nevertheless, it was an ideal setup for us. We assembled our fish receivers and anchored them out in the water, then quickly settled into the familiar routine of collecting by night and filling scuba tanks and feeding our fish by day. By now we knew every nook and crevice where fish would hole up, and it was not long before we had close to our desired number.

One day, during one of the frequent failures of the town's electrical power, the caretaker went out to refuel the generator. Instead of turning it off, he attempted to fill the tank on top of the still-running generator. You hear about these things but never expect someone to actually do it. Well, he did it. A little gasoline spilled on the hot generator, of course, and caught fire. He dropped the gas can and ran. Within seconds the entire generator shack was a raging inferno from the spilled five-gallon can of gasoline. We just stood and watched while the caretaker wrung his hands in mortification, some pain, and, I'm sure, embarrassment. The house was all right, but the generator was a charred ruin.

Oh, boy! This was something we definitely hadn't anticipated. There was nothing we could do to repair the damage, so when the boat came down from San Diego to pick up the fish we simply packed up and headed for home. I called the doctor when I got back to San Diego and gave him the bad news, whereupon he angrily asked how we planned to replace the generator and structure. Neither Sea World nor I felt that we were responsible. The doctor threatened to sue but quickly backed down when Sea World's lawyer responded that he would see him in court. I suspect he didn't want a spotlight on the tax scam he was getting away with down there.

Despite all the complicated logistics of organizing an expedition and the unexpected problems that cropped up, the days of those collecting trips to Baja California were some of the happiest of my life. Everything always seemed to come together. In a way it was like athletics, when a person's mind and body function as one and the end result is achieved without conscious thought. There were times later in my life, after I had taken up waterskiing, when I would feel one with the water and the ski. At other times, when I would consciously think about what I wanted to do, neither the end result nor the feeling was ever as good as when I was skiing "unconsciously," as it were.

As dissimilar as the three of us were, Bob Kiwala, Kelly McColloch, and I seemed to work well together. Much has been said in recent years by organization management trainers about the importance of teamwork, but teams can't be created by force—or by wishes. Effective teamwork seems just to happen when the right people come together with a common goal. Somehow, without trying, we had achieved what it is to be a "team."

IN THE DUMPS

Not all of our collecting expeditions were on such a large scale as those to Cabo San Lucas. One modest but memorable trip was by pickup truck to just south of San Felipe near the northern end of the Gulf of California. This was definitely a low-budget expedition (or as some put it, a real "Crummo Tour"), the object of which was to collect a few small fishes for one of the small exhibits in Sea World's marine aquarium building. We were after a number of species, but the main quarry was the juvenile Cortez angelfish. Like most angelfishes, the juvenile Cortez looks totally different from the adult—a strategy, apparently, that allows the young to live side by side with the adults without triggering territorial aggression.

I talked Bob Kiwala into joining me again, and he invited his artist friend Don Borthwick to come along. Don didn't collect or dive but wanted to draw and paint while Bob and I did our diving. We got a rather late start from San Diego and arrived in San Felipe shortly before sundown. After taking our time over supper in the last restaurant we would see for a while, we found it was quite dark when we were ready to head on to our campsite a few miles south.

San Felipe was at the end of the paved road, and from this point on it was dirt. Heading out of town we found what we assumed was the road south. The road immediately branched to the left, to the right, and then both ways at once. There was no moon, and we were beyond the last of the town's few lights. It was pitch black. We found ourselves going around in circles in this maze of branching dirt tracks. At one point, when we realized we'd passed the same point at least twice, we decided just to pull off somewhere for the night and figure out where we were in daylight. Taking one of the many choices of dirt roads, we soon stopped, lay our sleeping bags on the ground, and turned in.

Upon awakening in the morning we were a bit startled to see a burro browsing nearby and a couple of skinny yellow Mexican dogs scrounging for anything edible. We had pulled off and bedded down, it turned out, in the middle of the San Felipe garbage dump! Here and there among the profusion of beer cans and bottles were small piles of still-smoking rubbish. Thankfully there was nobody else around: we would have died of embarrassment if a San Felipe resident had seen three really stupid gringos camping in the town dump. We got out of there just as fast as we could.

That was certainly the low point of the trip. In daylight, by keeping the sea on our left and the mountains on our right, we managed to find the right road and headed south toward Puertecitos. A few miles later we saw a nice-looking cove, pulled off, and made our dives from shore. The water is always pretty murky near the north end of the Gulf, but Bob and I managed to collect most of the fish we were looking for at that one spot, and we transferred them safely to the holding tank in the pickup truck.

Heading back to San Diego, we stopped overnight at a house up in the mountains belonging to a friend of Don Borthwick's. We didn't expect it, but it got very cold that night at our elevation of four thousand feet. There was actually frost on the truck's windshield in the morning. More important, the temperature in the holding tank on the pickup had dropped to below fifty degrees Fahrenheit—though somehow all the fish were alive, even if they certainly weren't moving around with their usual vigor.

We should have known better than to stop on the top of the mountain, but it did demonstrate how the endemic fishes of the Gulf of California have evolved to withstand the extremes in temperature that frequently occur in the upper Gulf. In the summer, temperatures of eighty-five degrees are not uncommon, yet in winter when the cold wind comes down from the north the mercury can drop into the low fifties. Not many other fish in the world can take such a wide range of temperature.

BRINGING LIFE TO THE EXHIBITS

Ever since the days when I found that ordinary, nonbiologist-type people were interested in the marine creatures I kept in my home aquariums, I've enjoyed sharing with others the fascination I myself feel for what lives in the world of water. One obvious attraction is the beauty of these animals, but equally fascinating to me are the sometimes amazing strategies they have developed to survive and thrive in a potentially hostile world. In addition to bringing animals and plants into aquariums, I have tried to re-create the world in which they live.

When I arrived at Sea World in 1965, all the small aquariums had relatively sterile-looking artificial rock backgrounds—a far cry from the bottom of the ocean. True, the surfaces looked like actual rock, but unlike in the world underwater, nothing was growing on them.

One of the man-made habitats looked so unnatural that I decided to see if we couldn't bring back part of the real rocky bottom and install it in place of the artificial rock. After purchasing diver air-lift bags, three aquarists and I took our dive boat to Point Loma off San Diego, where we knew there were loose pieces of shale rock riddled with holes made by boring clams. The empty clam holes were now occupied by other animals, and the surface of the rock was totally covered with colorful sea anemones, solitary corals, and coralline algae. It is this profusion of growth in the underwater world that makes diving so fascinating, and

TIDE POOL TOUCH TANKS

The marine mammal shows, which Sea World widely advertised, were the primary draw for visitors. Personally, I was much more interested in other forms of life in the sea: animals with body designs and functions that are totally different from land-originated mammals like dolphins and seals. I found the amazing diversity of marine life far more interesting. I knew if visitors could see and experience some of these, they too would find them fascinating.

At one point when our daughters were young, we went on a car-camping trip from San Francisco to Vancouver, British Columbia. On the way we visited the Point Defiance Aquarium in Tacoma, Washington. I was most impressed with their aquariums filled with the invertebrate life of Puget Sound, collected by aquarium director Cecil Brosseau and his staff.

A couple of the exhibits in particular fascinated the visitors. They were nothing more than shallow, open-top, freestanding tanks about three feet in diameter placed in the middle of the public passageways. People could walk all around them and look down into the water. Beneath the surface were common invertebrates like sea stars, sea urchins, and sea cucumbers.

What was fascinating about these exhibits was that there was no barrier between the animals and the visitors. What's more, the animals were close enough to be touched. The most unique of these open tanks contained a good-sized giant Pacific octopus (*Enteroctopus dofleini*). This exhibit was for looking only, but what made it memorable was the close proximity of the octopus to the viewer. Remembering that early trip, I took Cecil Brosseau's simple but ingenious idea and expanded on it to create a major hands-on tide pool for the visitors to Sea World.

Designed to appeal both to adults and to the more inquisitive minds—and hands—of children, the exhibit had two components. One was a twenty-foot-in-diameter, donut-shaped pool with informational graphics and a staff member in the center to interpret the animal life. This pool, which was not for touching, contained a wide variety of intertidal and subtidal invertebrates and fishes, and its purpose was for looking and asking questions. The second component consisted of four shallow, dish-shaped pools, six feet in diameter and two feet off the ground. These four identical pools contained hardy species of sea stars, sea urchins, and a variety of seaweeds, all of which could be touched.

The height of the pools was designed for kids, which meant they were a bit awkward for adult-sized people—though that certainly didn't stop the big folks. It gave visitors an opportunity to actually touch another creature, to feel its texture and sense the seawater medium it lives in. The response of visitors to this exhibit was amazing.

The popularity of those pools has resulted in touch pools of various designs being included in aquariums worldwide. There is no question that the interactive nature of such an exhibit stimulates learning and creates a special appreciation of marine life in visitors. This awareness and appreciation should pay dividends down the line in protection of the environment and habitats of the animals represented. As the saying goes, we only protect what we care about.

Of course, this benefit is not without cost. When operated correctly, a touch exhibit is labor intensive, requiring the constant presence of a staff or volunteer interpretive guide. The guide's function is twofold: to interact with the visitors by providing information and answering questions; and to protect the animals from excessive or inappropriate handling. Even when the best of care is taken, some of the animals may die. And this raises some interesting points.

Animal rights activists object because they say the animals are "suffering," and some even believe all animals should always be free. But do animals lacking a central nervous system suffer? I don't think so; certainly not in the same way as mammals, with their highly developed nervous system. Should all animals be free, no matter what? Free, perhaps, to face death in an environment being polluted or destroyed by humans unaware of the damage we are doing? Better, I believe, that a few animals should die to save the many by teaching people that there are countless marvelous creatures living beneath the surface of the sea and that they, too, share this world with us.

---

this was what I wanted our aquarium visitors to experience. They might have seen pictures, but looking at a picture is not like seeing the real, living thing.

We anchored our boat over the shale reefs and dove down to look for rock slabs of an appropriate size to fit the exhibit. The rocks also had to be liftable because everything we were going to do would be done by our own muscle power. Maneuvering strong nylon netting under larger rock slabs and piling on as many smaller ones as would fit, we blew some air from our regulators into the five-hundred-pound-capacity lift bag that was tied to the netting—just enough to barely lift the rock-filled net off the bottom. It was then easy to push it toward—but not directly under—our boat forty feet above us. After a little more air was added, the bag, net, and rocks slowly started to rise to the surface. The air in the bag began to expand with the decreased pressure from shallower depth, and pretty soon the rig was rocketing upward at an alarming speed. I can only imagine what it looked like when it hit the surface.

That was the easy part. Now we had to manually lift the animal-covered rocks into the boat and lay them carefully on the deck. They were then covered with wet towels and sprayed with seawater for the trip back to the aquarium dock.

It was hard work, and on our next rock dive we found out that it could be dangerous as well. Fully loaded, one of the lift bags burst wide open just as it reached the surface. The rocks came crashing to the bottom below, just missing two aquarists who were busy gathering a second netful of rocks. That taught me never to trust a lift bag again; from then on we always swam well clear whenever a heavily loaded bag started up for the surface.

After the backbreaking work of fitting, jigsaw-puzzle-like, all the invertebrate-covered rocks into the aquarium tank, we added fishes like painted greenlings (*Oxylebius pictus*), rosy rockfish, and blackeye gobies (*Coryphopterus nicholsii*). The result was spectacular. It truly was like a piece of the bottom transplanted to the world above. None of the animals or plants were large, but their rich variety and vivid colors certainly captured visitors' attention. If only we could do this on a large scale, I remember thinking; wouldn't it be wonderful to give visitors the same sense of exploration and discovery that comes with actually being in the underwater world?

9

THE LURE OF SHARKS

THE YEAR WAS 1969 AND Sea World was doing well. However, the management knew that for the park's success to continue repeat visitors needed to be offered something new. With this in mind, a small group got together to brainstorm ideas that might be developed into future exhibits. The topic of sharks came up, stimulating considerable discussion and interest.

At that time, aquariums and oceanariums on the East Coast of the United States had successful exhibits of large, nearshore sharks like lemons (*Negaprion brevirostris*), bulls (*Carcharhinus leucas*), sandbars (*C. milberti*), and sand tigers (*Eugomphodus taurus*). Such hardy sharks didn't occur on the West Coast, however. Moreover, there had been little success in the past with the open-water pelagic sharks common off our deepwater Pacific coast. It was decided that these pelagic species might have potential as exhibit animals, but research needed to be done to find out for sure. To this end, Sea World committed financial support for a modest shark research program.

As a rule, shark species living close to shore tend to adapt to life in an aquarium better than those that live in the almost limitless water of the open ocean. These nearshore sharks, however—which on the West Coast include leopard sharks (*Triakis semifasciata*), horn sharks

A sleek, graceful blue shark. (Photo courtesy Marty Snyderman Productions)

(*Heterodontus francisci*), and swell sharks (*Cephaloscyllium ventriosum*)—are all bottom dwellers: not what the average aquarium visitor comes up with when visualizing a shark. The animals that everyone thinks of as "sharks" are the pelagic species: blues (*Prionace glauca*), makos (*Isurus oxyrhynchus*), the occasional smooth hammerhead (*Sphyrna zygaena*), and, of course, "Jaws" itself, the great white shark (*Carcharodon carcharias*)—to name the most common ones found off southern California.

In the late 1960s and early 1970s, though, marine biologists knew little about keeping pelagic sharks in captivity. Every one that had been brought to an aquarium had soon died. Obviously, something was not being done quite right. Sea World decided that a research effort might lead to success with either the blue shark or the mako shark.

As the curator of fishes, I took on the responsibility of designing this challenging research program. Going about it logically, I reasoned that three somewhat independent problems had to be solved, and steps followed, if a shark was to be displayed successfully in an aquarium. The first problem is the actual capture of the animal; second is the method used to transport the shark to the aquarium; and third is the design of the tank where the shark will live. Although these problems are somewhat independent, they are also linked. And of the three, the third is perhaps the most telling. For obviously, if you don't have a suitable holding tank ready for the shark to live in, you won't know if the first two steps were done right.

With no prior experience to help me out, I made my best guess and designed a relatively inexpensive, behind-the-scenes tank that I hoped would meet the needs of a pelagic creature like the blue shark. Trying to think like a shark—which is not easy, I found out—I reasoned that because blue sharks are rarely found near shore, they may be able to sense when they are approaching shallow water. They probably don't like sharp corners, since none exist in their watery world, and, because they swim constantly, they would most likely do best in as large a tank as possible. With those parameters in mind I designed a circular tank with a bottom that sloped up, becoming shallower toward the outer perimeter. Its size, fifty feet in diameter, was limited by how much money was available for the whole project. Bigger would obviously have been better, but you work with what you can get.

While the tank was under construction, we experimented with capture and transport methods. I had fished for blue sharks with rod and reel before from my own boat and knew from experience that they have a keen sense of smell and good vision. Our method of finding them was to take advantage of that sense of smell and allow them to find us. A fifty-pound burlap sack of ground mackerel was towed slowly behind the boat for half a mile or so, leaving a trail of odor that would smell like lunch to an almost-always-hungry blue shark.

For transport, we had a seven-feet-long-by-two-feet-wide plastic-lined tank made. It was restricted in size—but, I hoped, not too tight—because we were working out of a small, but fast, eighteen-foot outboard that didn't have much room or weight-carrying capacity. Knowing that the blue shark normally swims constantly in order to pass water over its gills, I made a flattened plastic mouthpiece that fit inside the animal's mouth. The mouthpiece had five holes on each side that theoretically lined up with the five gill slits of the shark. Through this de-

The first step in shark collecting: laying a delicious chum line of ground mackerel. (Photo by author)

vice, water supersaturated with pure oxygen was pumped by a small submersible pump sitting in the narrow transport tank.

Surprisingly, this improvised low-tech system worked and the blue sharks we practiced with were as lively as could be. We'd capture them and hold them in the long, narrow tank on board—facetiously nicknamed the "shark coffin"—for a couple of hours and then release them. We found, too, that they became quite docile and relaxed when held upside down and became active again when righted, and on being released they would take off swimming just like a normal, healthy shark.

The behind-the-scenes holding tank at Sea World was finally ready and the water system turned on. Now came our chance to see how well a blue shark would do. "Gator" Bill Ervin and I went out a couple of miles off Mission Bay (where Sea World is located), laid our chum line of delicious mackerel juice, and waited. Pretty soon a six-foot shark showed up off the stern. Using a heavy nylon hand line, I tossed out

A blue shark takes the bait and is quickly hauled on board by John Hart (*right*) and the author. (Photo © 1999 Sea World, Inc. All rights reserved. Reproduced by permission.)

the baited, barbless hook attached to a short wire leader. The shark quickly took the bait, and I instantly realized, as it nearly pulled me overboard, that this was no blue shark but most likely a much stronger mako. After finally getting it alongside our boat, Bill and I had an awful time lifting the heavy and uncooperative shark into the boat. Once it was lying upside down in the transport box, with the mouthpiece in place and the oxygen pump running, we took off for Sea World. I noticed that this particular shark repeatedly bit down on the mouthpiece, something I hadn't seen with the sharks we'd practiced with.

Hoisting the "shark coffin" off the boat and driving it the hundred yards to the waiting shark tank, we released the shark and it swam vigorously off. The transport method seemed to have worked fine. The shark looked good as it cruised around the fifty-foot tank.

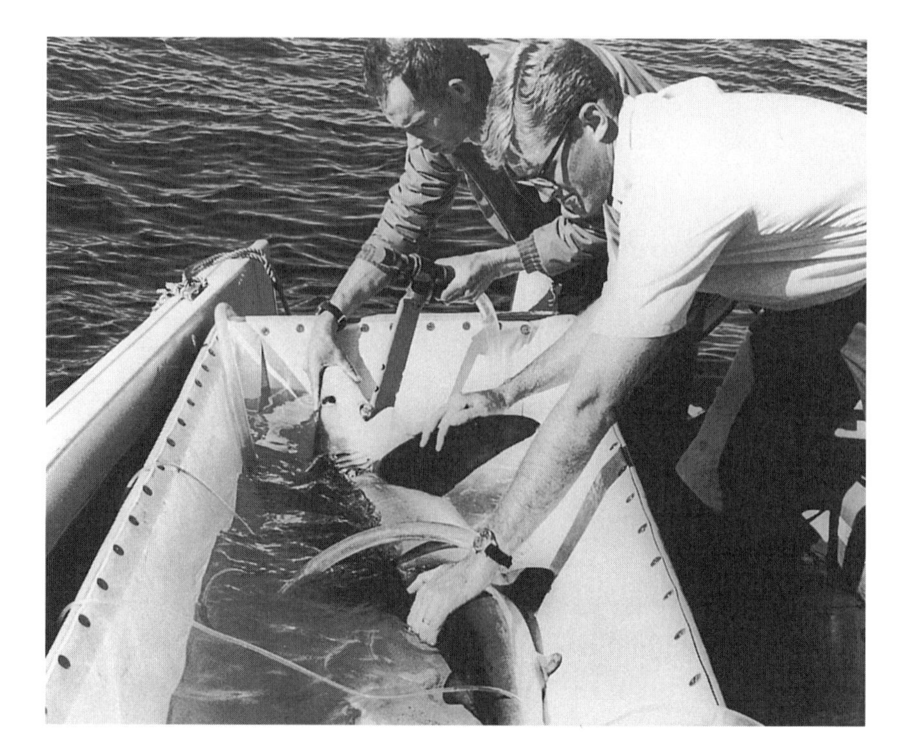

Lying on its back, the blue shark becomes still as life-sustaining oxygen is pumped across its gills through the plastic mouthpiece. (Photo

We spent the next eight days trying to get the shark to take food. It refused everything we offered. Only once, when we poured a bucket of mackerel blood into the water directly ahead of it, did it show any response; but it still wouldn't take the mackerel we dangled in front of it. On the eighth day it died, and only then did I realize that what we had caught was not a mako, but a young great white shark. I felt pretty stupid for not recognizing it when we caught it, but in the excitement of catching and getting the shark into the boat neither of us took the time to study its identifying features.

On the next trip out, and on many subsequent ones, we refined the collection and transport method to the point where we could bring two blue sharks in at the same time if they weren't much longer than six feet. Unlike the great white, the blue sharks readily began to feed

Back at Sea World, the shark box is lifted out of the boat for transfer to the experimental shark pool. (Photo by author)

and appeared to do fine—but only for a month or so. Then we noticed a change in their swimming posture: no longer perfectly horizontal in the water, they began to swim in a slightly tail-down position. Too, they seemed to be working harder at swimming. It became clear that they were losing weight, and autopsies later showed that much of the weight loss was from their liver.

Most species of pelagic, or free-swimming, sharks have large, oil-rich livers that, in addition to their metabolic functions, act as a buoyancy organ. As the oil in their liver was used up to provide energy they became heavier in the water and had to work harder to stay up. It became a vicious downward spiral: the harder they worked to stay up, the more they consumed of their liver, which caused them to grow heavier in the water, and so they had to work harder to stay up. In spite of being given all the food they could eat, they were using up their stored energy faster than they could replace it.

Our conclusion was that the pelagic blue shark is designed for long-distance cruising in a mostly straight line with a very low expenditure of energy and need for food. In its own environment, this design is very efficient, but I had created a tank that forced them to be constantly turning. Their metabolism was simply not designed for that much energy expenditure.

Our success with makos was almost zero but for different reasons. Makos are like the race car of the shark world: they need to actively swim at all times to stay alive. Just supplying them with lots of oxygen as we did with the blue sharks was not enough. After only forty-five minutes in transport they were barely alive when we arrived at the shark tank, and they died shortly afterward. Apparently the rhythmic contractions of the muscles during swimming play a vital role in the circulation of the blood of the mako shark.

Our lack of success caused us to terminate the research into local California sharks. Today, thirty years later, there still has been only mixed results displaying blue sharks. However, it is still possible they may do well in a much larger, correctly designed tank that has long straight runs and no turns except at the ends. Blues are one of the most beautiful and graceful of all the sharks, and it would be wonderful if they could be kept in good health in a large aquarium somewhere. Hopefully someone, somewhere, will have the opportunity to try it. Showing aquarium visitors the beauty of the blue shark may help stop the killing of hundreds of thousands every year for shark-fin soup, or as unwanted by-catch in the worldwide open-ocean longline and gill-net fisheries.

FLYING TEXAS SHARKS

With the end of our experimental work with local temperate-water sharks, we turned our attention to East Coast warm-water species that we knew did well in aquariums. The only problem was, they were three thousand miles away. How could we get them all the way to the West Coast? As luck would have it, an opportunity soon came along to test the feasibility of shipping tropical species of sharks by air.

Sea World was negotiating the trade of a pilot whale to the Searama oceanarium in Galveston, Texas. Searama had a number of bull sharks and lemon sharks in their large central tank that they had collected right in Galveston Bay, and they agreed to let me run a simulated shipping test with one of their six-foot bulls. The plan was to design and build a shipping container in San Diego and send it, together with life-support equipment, to Galveston for the test. When this was done I flew back there to set up the equipment at the side of their exhibit tank. The next step was to capture one of the bull sharks and simulate the conditions of a real shipment.

The main display tank at Searama was quite interesting. It contained a collection of just about everything from the Gulf of Mexico—alligator gars, big green morays, red drums, great barracudas, stingrays, giant jewfish, and, of course, several bull and lemon sharks. Young women in bikinis performed daily underwater feeding shows. I was amazed that these young women were swimming with notoriously dangerous bull sharks, which are known to attack more people worldwide than any other shark. And not only were these young women swimming with the sharks, but they were carrying food for the other fishes as well.

The bull shark lives close to shore and often goes into brackish water and sometimes even freshwater. It has been caught three thousand miles up the Amazon, and once up the Mississippi as far as Ohio. For many years this shark was known in other parts of the world by different names—the Zambezi River shark in South Africa, the Lake Nicaragua shark in Central America—until they were all shown to be the same species.

Curator Tom Whitman explained that it was quite safe to swim with the bulls because they treat the tank most of the time with a low dose of copper sulfate to control algae. This suppresses both the appetite and the aggressive nature of the sharks. Periodically they discontinue the copper, and the sharks are then fed. It still seemed risky to me, but they'd been doing it without a problem for a number of years, and I couldn't argue with that.

Tom organized the capture of our test bull shark. First his divers, using a large crowder net, herded the bull into a small connecting holding tank. It was then lifted out on a stretcher and lowered into my test transport tank. The oxygen was already on, and the bull shark quickly settled down under the mildly sedating effect of the high oxygen concentration. The test was to run for twenty hours, which was the estimated time for a real shipment of sharks from their tank in Galveston to our tank in San Diego.

Partway through the test a man named Gerrit Klay came by. He was an aquarist from the Cleveland Aquarium in Ohio and was down in Galveston to collect small bonnethead sharks (*Sphyrna tiburo*) to ship by air to his aquarium. The sixteen-or-so-inch bonnetheads, a relative of the larger hammerheads, were small enough that they fit into large plastic bags in standard Styrofoam shipping containers. He was very successful with this method, and it was quite an achievement to have bonnetheads exhibited at an inland aquarium in Ohio. Gerry Klay showed great interest in my experimental shipping test, especially the pump method of achieving high levels of oxygen. Two years later he left the Cleveland Aquarium to set up his own shark collecting business in the Florida Keys using the same transport methods he saw me using in Galveston.

After twenty hours the bull shark was released back into the exhibit. Although it was a little groggy for the first hour, it recovered completely and the simulated shipping test was deemed a success.

We next planned a real shipment of three bull sharks and one lemon shark from Galveston to San Diego. In addition, a whole menagerie of other fishes, including seventeen alligator gars, two huge three-hundred-pound jewfish, a beautiful giant green moray, and an assortment of smaller fishes from the Gulf of Mexico, would be transported—all in exchange for Sea World's one pilot whale. Of course, you don't just buy a ticket on a passenger plane for a shipment like this, so a Flying Tigers cargo jet was chartered to fly the pilot whale in one direction and all the fishy creatures in the other direction. The key people, besides myself, were Kym Murphy of Sea World, Tom Whitman of Searama, and—once again relenting to my plea for help—Bob Kiwala of Scripps. I'm sure he later regretted agreeing to come along "just for the fun of it."

All of our various-sized fiberglassed wooden shipping boxes had been sent ahead to Galveston, and at midnight we set about catching the animals in preparation for the drive to the Galveston airport and the waiting cargo plane. The four sharks were herded into the murky water of the small holding tank adjacent to the main display. To get them out we had to jump into the two-foot-deep pool and try to grab them as they swam by. It's quite unnerving having a six-foot bull shark push its way between your legs while you're trying to grab another. This was definitely in the days before government-mandated safety standards.

Somehow we managed to get all the animals into their respective shipping containers. The only casualty was one Searama diver who inadvertently backed up against the dive ladder in the main exhibit tank, where a totally unnerved giant green moray, taking refuge from the madness, had wound itself among the rungs. The moray lashed out in self-defense and bit the diver on the shoulder. He was rushed off to the hospital bleeding profusely but was okay the next day with a great story and some small wounds he could impress the girls with.

All the boxes of fish and the four sharks were finally loaded onto trucks. Because it was rush hour and we were running late, the City of Galveston provided a siren-screaming police motorcycle escort to the airport. They took their job seriously and were practically running motorists off the road so our trucks could pass. Eventually all the heavy boxes were on the plane and we were ready to go.

Concerned about the angle of the plane during takeoff, we asked the pilot to keep it as level as possible so water wouldn't spill out of the open boxes. We then strapped ourselves into the bucket seats at the rear of the cargo space behind our fish. The plane roared down the runway, and as soon as it was airborne the pilot pulled back on the controls and aimed for the sky.

Just as we had feared, water and fish poured out of the boxes. We couldn't do a thing about it until the plane leveled off a little. Then we scrambled around grabbing fish and tossing them back into their tanks. Kym Murphy was slipping and sliding on the deck rollers—intended for the easy moving of heavy freight containers—trying to pick up a slithering moray; somehow he got it back where it belonged. We then bucketed water from full tanks into ones that had lost water. The seawater that had surged out mysteriously disappeared down into whatever was below the deck.

The cockpit was open and Bob went up to talk to the pilot. He said, "That was a pretty impressive takeoff. You lost two hundred gallons

After a long flight, this lemon shark is recovering in the holding tank. (Photo by author)

of saltwater!" Without turning around the pilot said, "You take care of the fish and we'll take care of the airplane." Well, okay, we thought. Glad it's your plane, not ours.

All went well until we crossed over west Texas and hit a violent thunderstorm. The plane was thrown around and the groupers and alligator gars freaked out, leaping and splashing and spilling even more seawater down below. Moving about to check on the pumps, batteries, and our fishes was extremely difficult thanks to the steel deck rollers, which made walking upright virtually impossible. So we crawled around on hands and knees, hanging on to whatever we could grab.

Things calmed down after Texas, and pretty soon we were about an hour out of San Diego. About then all the lights in the plane went out: we instantly knew that the spilled seawater had shorted out the electrical system. Now it was pitch black in the back except for the meager beams of our flashlights. We all prayed that it was only the lights and not the plane's controls that had shorted. Up in the cockpit red lights were flashing as the engineer peered into the electrical panel, a concerned look on his face.

When we got close to the San Diego airport, the pilots had a problem getting the landing gear down. We circled around for a while before they succeeded. At this point smart-ass Bob went back up to the cockpit and said, "Well, we took care of the fish; how you doing with the airplane?" He got no response. We'd been told that the plane was supposed to leave that night for the Philippines, but we heard later that the plane was laid up for two months for electrical system repair. That flight, together with another later one by Gerry Klay, led to strict design and construction guidelines of shark-shipping boxes so they couldn't lose water. From years at aquariums experiencing what seawater can do to electrical systems, I knew that was a wise policy. I don't think they addressed problems caused by know-it-all pilots, however. Flying Tigers refused to charter planes to Sea World for some years after, and Gerry Klay was banned by all cargo airlines for a while.

In spite of the difficulties, the sharks, the groupers, and all the smaller fishes made it. Sadly, we lost some of the seventeen gars because of overcrowding. Unlike most fish, gars are air breathers, but there was so little room in their boxes that they couldn't all get up to breathe when we hit that turbulence over Texas. Indeed, we learned a lot from that ambitious, pioneering shipment.

A RECONNAISSANCE TRIP

Our success with the tropical sharks from Galveston encouraged us to look for a closer source of warm-water species. The Gulf of California isn't far from San Diego, but little was known about the abundance and distribution of the sharks in the upper part of the Gulf.

There was one useful publication on eastern Pacific sharks, written by Susumu Kato of the U.S. Fish and Wildlife Service, but none of us at Sea World had any personal shark experience in the area. No cargo planes fly down to the upper Gulf, and the only paved road went as far as San Felipe in Baja California. We decided to make a reconnaissance trip and put out baited setlines to see what we could catch. We didn't plan to bring any sharks back on this first trip; we just wanted to find out what was there.

Because of its long, narrow shape, the Gulf of California at its northern end, near San Felipe, is noted for having the second highest tidal change in the world after the Bay of Fundy in Nova Scotia. During the time of the new and full moon spring tides, the water level rises twenty-five feet from low to high tide in six hours. That's an incredible amount of surging water.

The local fishermen of San Felipe have learned to take advantage of this tidal fluctuation. When a fishing boat needs work done below the waterline, they simply drive the boat at high tide to a low spot near the middle of town and wait for the tide to go out. The boat is then high and dry and they have a few hours to do their work before the water returns to refloat their boat.

Our team consisted of two Sea World aquarists, John Hart and Jerry Kinmont, and myself. Together we loaded up John's truck and then towed our eighteen-foot outboard collecting boat down to San Felipe. Our plan was to use it to set out small, bottom-fishing setlines. We had also made up a half-mile-long floating setline that we would set from a bigger boat to see what larger sharks might be there. This line was to be buoyed up by a series of truck inner tubes, with the baited hooks hanging twenty or thirty feet below the surface. An anchor at each end would keep the setline from drifting or being towed away by the current—or by any creature that might get caught.

We chartered a small fishing boat and hired its owner to take us out a couple of miles from shore to put out our gear. We'd brought boxes of frozen mackerel to use as bait. The baiting and setting of the longline went without mishap, and we returned to San Felipe Harbor to while away a few hours. When we headed back out to see what we'd caught, there wasn't a trace of the half-mile-long line with its twenty inflated inner tubes. We circled around for some time looking, but eventually gave up. We decided to return in the morning and search again before writing the line off as lost.

This was the first day, and it wasn't looking like a good start to our trip. We were totally baffled as to how the line could have disappeared

Sunset on the barren desert road to San Felipe: a most unlikely-looking place to catch a shark. (Photo by author)

without a trace. We had visions of some huge shark towing it off, anchors, floats, and all. The next morning we got up early and headed out to the same area—and there were all the inner tubes bobbing quietly at the surface right where we had left them. Suddenly it dawned on us: the tremendous tidal change also causes powerful currents. The strong current had sucked every inner tube completely underwater and out of sight. The pull on the anchors from that ripping current must have been tremendous, but they dug in and held.

We began to pull in our line to see if we had caught anything. Soon we felt something large on one of the leaders, and it was trying to head away from the boat. Eventually, as we continued to pull in, a head and large mouth came into view through the murky water. We were excited to recognize it as a still-alive, ten-foot-long great white shark. Unfortunately, even though it was still swimming weakly, the shark was brain-dead from lack of oxygen. Confined too long on the line, it had been unable to swim freely and to pass enough life-sustaining water

Our crippled boat is towed back to San Felipe by helpful Mexican fishermen. (Photo by author)

across its gills to maintain its oxygen-sensitive nervous system. I knew from past experience that it wouldn't survive if we released it.

The owner of the boat said he wanted to keep it to sell in the fish market, but at an estimated eight hundred pounds it was much too heavy for us to lift into the boat. So he killed it with a knife lashed to a net handle and we tied it off to the side of the boat. Pulling in the remaining longline, we felt another shark. It also turned out to be a great white, this time an eight-footer, which we managed to get on board. White sharks make for good eating; this would be a fine catch to bring back to the village.

Our first day had been most strange: first a magically vanishing and reappearing line, and then two unexpected great white sharks. Not what we were after, but still very interesting.

The skipper, not knowing what to do about the larger shark alongside, decided to turn it loose tied to one of our inner tubes and come back for it later with more help. I was a bit puzzled by this decision. Didn't he know that in a few hours the tremendous currents that surge up and down the Gulf every time the tide changes would carry his shark and its float miles away? I was sure he'd never see the fish or the tube again.

We planned to do more fishing with the smaller longlines and were just rounding the north point when our 150-horsepower Mercury outboard made a screeching sound like tortured metal and shuddered to a stop. It was completely seized up. What now? We were about three miles from San Felipe in a heavy boat with one locked-up engine and two little paddles. A slight breeze was blowing south toward town, though, so we optimistically hung the vinyl shark stretcher from the shark-lifting davits to make a crude sail, and soon we were sailing along at a barely perceptible speed. Of course, if the current had started to run north, we would have gone backward at a much higher pace.

We had plenty of drinking water on board—a wise precaution in these unpredictable waters—so we just lay back and relaxed. A couple of hours went by and along came a Mexican fisherman in his trusty panga, one of the sturdy, efficient, seaworthy fishing skiffs used all over Mexico. A 50-horsepower Johnson outboard planes these impressive little boats along at a very respectable speed and fuel economy, regardless of their load. Laughing at our predicament and our huge nonfunctional engine, the fisherman kindly gave us a tow back in to San Felipe.

That pretty much ended the reconnaissance trip for sharks: there wasn't much we could do with an inoperative boat. Still, despite the brevity of the trip, we had learned something. Because we'd caught white sharks, which like cool water, we concluded that this was the wrong time of year for the tropical sharks we wanted. We had chosen this time of year to avoid the problems that scorching hot weather would have caused if we'd tried to ship live sharks up to San Diego. But clearly that didn't help if our quarry was basking in the warmer waters of southern Mexico.

LOSING SEA WORLD'S BOAT

We decided to resume our shark hunt in the early fall, when the water would still be quite warm but the air would not be a hundred-plus degrees like it is in the middle of summer. Even though our reconnaissance trip had been cut short, the plan for our second expedition was to bring live sharks back to Sea World. We had collecting permits from Mexico City—which, we hoped, meant there would be no trouble at the border or with the Oficina de Pesca official in San Felipe. Because we planned to bring live sharks back with us, the arrangements for this trip were much more elaborate and would, among other things, involve setting up a holding tank on the beach to keep the sharks in prior to driving them out.

I was again working with John Hart and Jerry Kinmont, as well as veterinarian Jay Sweeney and Dr. Murray Dailey, a parasitologist from Long Beach State University whose special interest was the parasites of elasmobranchs—sharks and their relatives. He was most eager to check out the internal parasitic fauna found in sharks and rays from this seldom-studied region.

We towed down the same eighteen-foot Thunderbird collecting boat, now equipped with a new 150-horsepower Mercury outboard engine. Arriving in San Felipe, we checked into a small motel on the beach at the south end of town. Murray Dailey was really impatient to get at his favorite animals—shark parasites—and he begged me to let him put out a small setline to see what he could catch overnight. Giving in to his pleading, I helped him launch the boat. Murray then laid out the small baited setline not far from shore and, when he was through, anchored the boat off the beach in front of the motel and swam in.

We all went into the center of town for supper and to discuss our plans for the next day. Arriving back at the motel after dark, we sat on the patio looking out to sea. Suddenly someone said, "Where's the boat?" We stared out into the dark; none of us could see it. We knew there was enough light shining out from the motel to illuminate the light-colored hull, but it simply wasn't there! We also noticed that an offshore wind had sprung up and was blowing out to sea.

A feeling of panic overcame us. We all jumped in the truck and drove into town to see if we could find a panga fisherman to run out and try to find the boat. We managed to raise one sleeping fisherman from his warm bed. Muttering something unintelligible in Spanish, he headed down to his boat and out into the dark sea. He came back in an hour, said, "No good, too dark, too rough, we go look *mañana,*" and then went back to bed.

The offshore wind had picked up considerably by now, and we knew the boat must have drifted a good distance from shore, on its way toward the mainland of Mexico, seventy miles across the Gulf. We mentally reconstructed what had happened. Being unfamiliar with the extreme tides in San Felipe, Murray had put out what he thought was an ample length of anchor line. But it was low tide. When the tide came back in, the rising boat simply lifted the anchor out of the sand, and away it went with the offshore wind. It really wasn't Murray's fault, but he felt terrible. I blamed myself for not warning him about the Gulf's tremendous tides.

Feeling totally helpless, we went back to the bar for some tequilas and desperately tried to think of some way to find Sea World's boat. There is a little dirt air strip in San Felipe; maybe if we could talk an American pilot into flying out over the Gulf, he'd be able to spot the little boat drifting along.

After a sleepless night we got up at dawn to drive to the airfield. By now the wind was blowing about forty knots straight out to sea. But luck seemed to be with us: a Cessna was just getting ready to leave. We told the pilot of our predicament and asked if he would help us by taking a quick look for our missing boat. He said he was sorry but if he didn't get out of there right now and head for the States he wouldn't be able to take off later if the wind got any worse. We watched him taxi to the end of the strip, take off, and disappear. Our last hope of getting the boat back was dashed.

Even though it was still early morning, we headed for the bar, where we sat staring blankly out at the choppy sea, mentally preparing our résumés for our next jobs. Three tequilas later Murray said, "Look! What's that?" and pointed to two specks heading toward us from the horizon. As they came closer we recognized the distinctive Sea World Thunderbird and alongside it a panga, with one man in each boat.

Grabbing a large amount of cash from my room, we raced down to the harbor to greet the two returning boats. I could have kissed the two fishermen! What they were doing way out there in this awful wind I never did find out, but they had spotted our boat sailing merrily along toward the Mexican mainland. The keys, of course, were in the ignition and it had a full tank of gas. One of the men had just hopped in, started the boat up, and headed back to San Felipe, with the panga planing along beside it.

That was the best two hundred dollars of someone else's money I ever gave away! The fishermen were delighted to get this unexpected windfall from the crazy gringos. I was delighted to keep my job and happy that Murray Dailey didn't have to live with the lost boat on his conscience. I never told Sea World management what really happened and managed to conceal the missing money through some creative bookkeeping supported by indecipherable Mexican receipts.

SHARKS IN MEXICO

With that nerve-wracking event behind us we turned our attention to catching sharks. The first project was to set up a twenty-foot-diameter circular portable plastic swimming pool on the beach. Having already experienced the extreme San Felipe tides, we wanted to be absolutely sure it was well above high tide. The logical approach was to ask the people who lived right there at the motel. They came out and pointed to a spot where it would be perfectly safe from the highest possible tide. Using buried plywood boards as sand anchors, we laid out the pool and its sunshade, an inexpensive army surplus parachute that worked well and was surprisingly windproof.

The full moon came a week later, and with it the spring tide with its extreme highs and lows. We watched as each day the high tide crept closer and closer to our precious shark pool. One day it was lapping at the base of the pool, and we hadn't even reached the maximum tide. I could have killed the man who said the pool would be completely safe, but he made sure he was nowhere to be found.

Faced with two more days of high tides, we rallied our meager forces and built a seawall between the pool and the incoming tide, using our empty fiberglass shark transport boxes and filling them with sand. Guests staying at the motel pitched in and helped shovel sand to fight the relentless ocean. We looked and felt truly stupid for erecting our pool where the ocean could reach it; however, we fought on and even-

All hands battle the high tide that threatens to wash away the shark pool at San Felipe. (Photo by author)

tually passed the peak of the high tides. Now we had two clear weeks before the new moon and its series of high tides came.

Collecting began. We put out longlines in the bay of San Felipe; we caught mostly rays of several species and a few small sharpnose sharks (*Rhizoprionodon longurio*). Outside the bay in deeper water we caught a huge four-hundred-pound mako and some four-foot-long bignose sharks (*Carcharhinus altimus*), which were listed as rare in Sus Kato's shark book. They were small, but they did well in our holding tank.

One of the locals told us about a *tiburoneros'*, or shark fishermen's, camp around Punta Smith a few miles south; we decided it would be smart to visit the people who make their living catching sharks. The camp was inaccessible by road, so we took off in the T-bird and headed south. Arriving in the afternoon, we were greeted warmly by the fishermen, who told us they had set their nets but wouldn't be pulling them until after dark. That was okay by us; we would wait. By now we'd been around San Felipe long enough to know our way back after dark.

In the late afternoon they graciously invited us to join them and their families for a supper of fish tacos—made, of course, from dried shark meat. Next to their small hut were improvised drying racks for strips of shark meat that would be shipped to the mainland of Mexico. The shark fins would be shipped to Asia. The smell of the drying shark meat was almost overwhelming, but the tacos were surprisingly good, maybe because we hadn't planned on staying out so long and we were starving by then.

Nightfall came and the tiburoneros set out in their panga to pull the nets they had set not far from shore. It was one of those beautiful calm, warm evenings that can be so magical in the Sea of Cortez. Not a breath of wind could be felt, and the sounds of the fishermen singing and dancing on the bottom of their boat came to us clearly across the mirror-smooth water as they pulled their nets. Hearing such happy sounds, we looked forward to seeing what kinds of sharks they had caught. They returned to camp and told us there had been nothing in the nets.

That evening with the hospitable tiburoneros still haunts me. We knew that even when fishing was good they were very poor and to catch nothing must hurt, yet there they were singing and dancing as they worked, knowing all along that they might end up with nothing. What was the secret to their happiness? We wealthy Northerners, with all our material possessions, would have been cursing our bad luck if we'd had a night like they had.

This collecting trip was not a great success. We ended up with two rather small bignose sharks that we brought back to Sea World. We'd had three beautiful two-foot-long hammerheads in our holding pool, but some kids saw them as great things to play with and killed them. I'm sure they didn't mean to hurt them when they were grabbing their tails and dorsal fins; they just didn't know how delicate they were.

Because of the poor results and the difficult logistics, this was our last attempt to collect sharks in the Sea of Cortez. Although expensive, it turned out to be more practical to collect on the East Coast of the United States and transport the sharks across the country.

10

TANNER BANK AND MEXICO EXPO

A SPECTACULAR DIVE

Although my day-to-day Sea World duties took up most of my time, I jumped at every opportunity for a diving adventure. One of the highlights of my diving experiences was an offshore submerged bank called Tanner Bank, which, some hundred miles west of San Diego, rises in one spot to about thirty feet from the surface. The group of divers on this trip consisted of Bob Kiwala, underwater photographer Chuck Nicklin, Sea World's assistant curator Ray "Chub" Keyes, and myself.

We were unable to locate the small thirty-foot-deep pinnacle, so we anchored somewhere near it in eighty feet of water. As it often is that far from shore, the ocean was clear and blue, and, looking over the side of the boat, we could clearly see the rocky bottom. There was a pretty good current running, which meant it took real effort to swim from the stern to the bow so we could go down the anchor line. The line would also be our visual reference for getting back to the boat at the end of the dive. You don't want to come to the surface astern of the boat out there; it could be your last dive if the current was too strong to swim against and the crew didn't see you come up.

Once in the water I was enthralled by its almost unbelievable clarity. The sun was overhead, and I could see two other divers well over a hundred feet away as clearly as if they had been next to me. It was the clearest water I have ever seen in any ocean.

On the bottom everything had a beautiful purplish hue. Even fishes like the painted greenling that as a rule are a mottled reddish color were purplish. There were lots of rosy rockfish, which are normally quite beautiful, but here they were truly striking with lavender markings on their bright orange bodies. I also saw numerous smallish clumps of the purple hydrocoral.

What struck me also was how clean everything was. There was none of the sediment that you generally find even on Catalina Island. During storms, this reef must get violently swept by the open ocean swells that come from far away in the Pacific Ocean, as well as washed by the prevailing coastal currents. We were fortunate that this day there was almost no swell to generate underwater surge, which makes diving and especially collecting difficult. It was calm, and we enjoyed the beauty of this unique place at our leisure.

I never made it back to Tanner. It's probably just as well, as it's unlikely that a second trip would have had the perfect conditions we experienced that day. Tanner Bank is forever imprinted in my memory as truly special.

Weather conditions can make such a difference in a dive. San Nicolas Island, the farthest out of the Channel Islands, is supposed to be a wonderful place to dive, but on my one and only trip there a twelve-foot swell was running. This made the island impossible to dive except for one spot in its lee, and even there the surge was so bad that visibility was three feet and we were tossed around like wet rags. Without question, it was a long way to go for a totally frustrating diving experience.

EXPOSITION IN MEXICO CITY

As part of the process of granting permits to collect from the Gulf of California, the Mexican government, perhaps understandably, always expected something in exchange. One year we hosted marine biology students from the University of Mexico City, who spent a couple of months learning the art and science of aquarium-keeping and accompanying us on our collecting trips. We always provided the university with copies of underwater photographs and films taken on our expeditions as well as preserved specimens for its scientific collections. And one year I set up a refrigerated seawater system in the university's biology lab.

In 1971 we again applied for a permit. This time the Mexican government had a specific request: we were expected to participate in a major fisheries exposition scheduled for the Christmas holidays. It was to be held in the fairgrounds in Mexico City and would last five weeks. After asking if they had anything in particular in mind, we were told that, among other things, they wanted sea lions, harbor seals, and one adult male elephant seal. They particularly insisted on the elephant seal because it had been collected by Sea World under a previous permit from Mexico's Isla Guadalupe, off the west coast of Baja California. In addition to the marine mammals, they wanted a touch pool and a tropical aquarium full of Mexican fishes from the Sea of Cortez, together with informational graphics in Spanish about the animals.

Try as we might, there was no budging them on these demands. Most of what they wanted was certainly possible, but the very thought of transporting a two-thousand-pound elephant seal from San Diego to Mexico City and back just for a temporary exhibition was scary, to say the least.

To add to our worries, the organizers then told us that no preparatory work could be done at the fairgrounds until three weeks before the expo was to open; a German industrial equipment exposition would be there before us, and construction was impossible until it closed. Still, they repeatedly assured us that it was no problem to get us whatever we needed, that everything we asked for would be ready. We told them that the elephant seal would need a very large pool of seawater and the seals and sea lions a slightly smaller one.

Because excavating the fairgrounds was not permitted, both of these pools would have to be built above ground, with raised viewing areas around them for the people to see the animals. Again they calmly told us that everything we asked for would be there. I said that the tropical aquarium and all the graphic panels about the animals would have to be inside a building to protect them from the sun. "*No hay problema.*"

We really had no choice but to agree. In our hearts we just knew that some kind of disaster lay ahead—a dying elephant seal with full media coverage being the vision we feared the most.

I thanked my stars that I didn't work with mammals, and I started my own preparations for the touch pool, fish exhibit, and informational graphics. The tropical aquarium was a nice-looking four-foot acrylic cube, with its life support system beneath it. The touch tank, ten feet long and irregularly shaped, was fabricated out of fiberglass simulated rock. Given the time we had, the rock didn't look especially realistic, but at least it held water. It would have an under-gravel filter in the bottom for biological filtration and a refrigeration unit to keep it cold.

Among the countless problems we encountered was the fifty-cycle electricity available in Mexico City; for that, special pumps and refrigeration were purchased. The biggest problem, however, was the biological filtration, which had to be fully functional from the very first day if the fishes and invertebrates were to stay alive and healthy. The normal procedure during the set-up of a new aquarium is to seed the gravel in the new tank with bacteria—which break down waste products into a nontoxic form—from a well-established aquarium and then to feed the bacteria nitrogen to get them to multiply. This process generally takes three or four weeks, but I would only have a few days to set up the aquarium systems and put the fish in just before the exposition opened to the public. Introducing a lot of fish to an aquarium that doesn't have a functioning biological filter is a sure way to kill them. We couldn't run the risk of that happening.

The solution was to air-freight biologically active filter sand from San Diego to Mexico, arranging for it to arrive the day after the aquarium had been filled with artificial seawater. For this plan to succeed, the billions of bacteria coating the grains of sand had to be kept alive and healthy. The sand would therefore be shipped moist but without liquid water, and the plastic bags containing the sand would be filled with oxygen. I have no idea if this had ever been done before, but it was the only chance I saw of having live, healthy fish in the aquarium.

Everything went according to plan. I flew to Mexico City with the touch pool, the acrylic aquarium, and all the assorted life-support equipment. Arriving at the fairgrounds, I was amazed. The aquarium building was complete, and the two large concrete pools for the seals and sea lions were filled and ready to go. Bleachers and ramps had been built around the mammal pools to give the visitors good views of the animals. Everything was painted and functioning, just as they'd told us it would be. Meanwhile, frantic activity was going on all around as fisheries exhibitor's booths were installed and stocked.

The fairground crews had been responsible for all of the construction, and I was most impressed with both its quality and their speed. We were told that the crews had worked night and day, and seeing what they had achieved I could well believe it. They had done a remarkable job.

I set up the aquarium and the touch pool and filled them with artificial seawater made from salts shipped from San Diego. The next day, right on schedule, the five hundred pounds of moist, oxygenated filter sand arrived; it was immediately poured into the two tanks and the filter pumps were turned on. The following day Styrofoam boxes of live fish and invertebrates arrived by plane from San Diego, to be installed in their appropriate exhibits. All went according to plan, and, most important, the crucial filter bacteria got right to work taking care of the fish's wastes. My part of the show was ready.

The marine mammals, which were under the care of Sea World's veterinarian, Jay Sweeney, arrived on a chartered DC-8 jet freighter. The harbor seals and sea lions were no problem to move but the crated elephant seal, weighing in at a ton, required a crane to be lifted off the truck and onto the dry area of the pool. Released from its shipping cage, it lumbered eagerly into the water with a huge splash. It was an impressive sight.

The exposition opened to the public two days later, and people came in droves. Sea World's marine mammals and aquariums were tremendous draws, having been heavily advertised throughout Mexico. Visitors stood in line for hours. The security staff, in an effort to keep things moving, allowed each person less than five minutes inside the aquarium building. That amounted to walking in the door, slowly walking by the aquarium and graphic exhibits, and exiting out the back.

I got the sense from watching all this that these people were starved for contact with nature. Many, if not most, in this giant inland city had probably never seen the sea or the creatures that live in it. They may have had television sets, but even if they watched nature shows on TV, it's not at all the same as seeing the real live animals.

I believe there is an innate need in humans to have some form of contact with other living beings. This need may explain the popularity of zoos and aquariums in the cities as well as the popularity of keeping animals as pets and even growing potted plants in our homes.

11

THE REVILLAGIGEDO ISLANDS

OUR DECISION TO STEER clear of collecting at Cabo San Lucas for a while, not to mention the ill-fated collecting trip to Los Frailes, led us to look elsewhere for fish. An opportunity arose to make a collecting trip to Isla Socorro, the largest in the group of four islands called the Revillagigedos (pronounced *ray-vee-ya-hee-hay-dos*), located 250 miles south of the Cape. The trip would be aboard the brand-new sixty-five-foot sportfisher *El Navegante,* skippered by Tom Beach, the nephew of industrialist Leonard Friedman.

We jumped at the chance. Not only was the area virgin territory for diving, so far as we knew, but it was also home to the beautiful, bright-orange Clarion angelfish (*Holacanthus clarionensis*), named for Isla Clarión, the second largest in the archipelago. Long-range sportfishing boats had been going to the Revillagigedos for a few years and returning with great catches of game fish. They also came back with stories of the numbers of game fish lost to sharks, adding that we'd be crazy to dive there. Their tales were a bit unnerving, but we weren't about to pass up such an opportunity.

We explained to Tom the problem we'd experienced with the cold water off northern Baja California killing some of the tropical fish we were transporting back to San Diego. He said he would be happy to install a seawater heating system using engine manifold heat; his uncle would even be willing to pay for the modifications. If he would do all that, we were more than happy to go to Socorro with him.

Again I was successful in talking Bob Kiwala into coming along, and he invited Ron McConnaughey, also from Scripps Institution of Oceanography. They wanted to make a scientific fish collection—in the form of preserved specimens—at Alijos Rocks (pronounced *ah-lee-hos*), lonely, uninhabited rocky pinnacles 180 miles off the west coast of Baja California. The rocks, actually the tips of a submerged oceanic volcano, are located two-thirds of the way down the peninsula at the juncture of two biological zones, the temperate to the north and the tropical to the south. Because of the remote location, no one had ever made a scientific collection there. In fact, though the area had been fished by hook and line, it was considered rather unproductive—mainly, we soon learned, because of the large number of sharks that attacked almost every fish they hooked up.

Now, coming from scientists at Scripps, that information began to worry us. The fish collection we planned was to be made using scuba and spreading an ichthyocide called rotenone, which kills fish relatively quickly by blocking their oxygen uptake. After the rotenone was spread, we would dive down, gather up all the dead fish we could, and stuff them into net bags to bring back up to the boat. The fish would then be preserved for later classification at Scripps.

ARMED WITH BANG STICKS

Shark stories put temporarily aside, the trip was on, and we loaded the boat with our usual collecting equipment plus several gallons of rotenone and a number of new bang sticks we'd invested in. A bang stick is a defensive device a diver carries. When it is thrust against a threatening shark, the contact will detonate a rifle, shotgun, or handgun cartridge. Unlike firearms used above water, bang sticks have no bullets; it is the explosion that incapacitates the shark, rather than a projectile. Reputed to be deadly, they come in a variety of styles and calibers. None of us had actually used one before, but they made us feel somewhat safer. Now all we had to worry about was our dive buddies nervously waving their bang sticks around.

In addition to Bob and Ron, the team included underwater photographer Chuck Nicklin, aquarists Kym Murphy and John "Muggs" Carlton from Sea World, Larry Casteñares from the University of Mexico City, and me. We departed San Diego and two days later at daybreak arrived at our first stop, the bleak, lonely Alijos Rocks. The weather was gloomy and overcast. Even without the thought of sharks, none of us felt like going into that gray water. Under the best of conditions, there seems to be a distinct lack of enthusiasm for very early morning dives when you have just crawled out of a warm bed.

We anchored at a spot where there was about sixty feet of water below the boat. Bob and Ron went in first, carrying their bang sticks, to spread the rotenone. Shortly afterward the rest of us would join them to gather the stunned and dying fish. Among us we had four bang sticks in case we were harassed by the aggressive sharks we anticipated meeting. Down we went, our heads swiveling in every direction to look for the sharks we knew would be drawn to the hundreds of struggling, dying fish.

After all our preparations—and worry—not one shark was to be seen. In a way, it was a letdown. Here we had psyched ourselves up to meet hordes of sharks, and they just weren't there. We set to work gathering as many fish as we could of all sizes, right down to the tiniest gobies drifting out of the crevices. Dick Rosenblatt, the ichthyologist at Scripps for whom the scientific collection was being made, had expectations of finding new species at these remote islands; if there were any, they would be the small, geographically isolated, resident fishes that, over time, had evolved to become genetically different from their ancestors on the peninsula of Baja California.

We finally figured out why there were no sharks: it was late February, and the water was still too cool. We never learned which species of sharks had caused all the trouble for the hook-and-line fishermen from Scripps, but I suspect they were tropical and subtropical species like Galápagos sharks (*Carcharhinus galapagensis*), blacktips (*C. limbatus*), and hammerheads. When the water cools down in winter, they all head for southern Mexico or Central America, where the temperature remains warm enough for them throughout the winter. A few months later it would have been, as they say, Shark City around Alijos Rocks, and our dive would have been a lot more exciting. Nevertheless, I noticed that none of us was complaining.

With the fish collection accomplished and the specimens preserved in formalin, we were eager to get to the live fish at the Revillagigedos. First we had to refuel and top off the boat's freshwater tanks at Cabo San Lucas, but soon enough we were on our way—a 225-mile, twenty-five-hour run due south. The weather was good, and it was a comfortable ride downhill with the prevailing northwest swells on our tail.

The first island we came to was San Benedicto. This small, uninhabited, and recently active volcanic island was completely covered in ash and lava, a massive eruption in 1952 having wiped out all visible life. Since then very little vegetation had managed to gain even a precarious foothold on the barren lava. The water surrounding this northernmost of the Revillagigedo Islands was home to giant manta rays (*Manta brevirostris*), which later observations indicated might be permanent residents. This is also where a world-record yellowfin tuna (*Thunnus albacares*) was taken in 1977—a giant powerhouse weighing 388 pounds.

We made one dive here, but it wasn't good terrain for collecting the kind of fish we were after. Piles of giant volcanic boulders allowed them too many escape routes from our hand nets.

We continued on to our primary destination, Isla Socorro. Although it, too, was volcanically active, it had not erupted in many years; still, wisps of smoke or steam could be seen coming from the island top.

The skipper dropped the hook in about thirty feet of water in a lovely, protected cove. Eager to see what we might catch here, we fished over the side of the boat, and our baited hooks were immediately attacked by a horde of beautiful crosshatch triggerfish (*Xantichthys mento*) that acted like starving piranhas. Their tiny mouths and sharp teeth stripped our hooks clean of bait in seconds.

We jumped in the water and snorkeled around, amazed to see dozens of the "rare" Clarion angelfish. We also spotted a few small Galápagos sharks, two to three feet long, and an occasional small silvertip shark (*Carcharhinus albimarginatus*). After checking us out, they swam on about their business. That made us feel a little less nervous about our

El Navegante safely at anchor in a Socorro Island cove. (Photo by author)

first night dive, which we planned to do after supper. Even though we hadn't seen anything that posed a real threat, the fishermen's stories about the ravenous sharks of Isla Socorro were fresh in our minds.

Supper was over and all signs of daylight were long gone. We couldn't procrastinate any longer, so slowly we began to get our dive gear together. We'd decided to dive in two groups of three: while two divers would work together on the bottom collecting fish, the third would hover just above them with a loaded bang stick for shark protection. We were quite nervous. Careful to avoid any shark-attracting splashes, we slipped quietly into the black water off the swim step. No sharks were in sight—at least, we saw none in the narrow beam from our dive lights. So down to the bottom we went.

Trying to work efficiently and calmly, we hand-netted fish and transferred them into our plastic bags as quickly as possible so the vibrations from their struggling didn't sound the dinner bell for hungry predators. The third diver, riding shotgun above us, kept flashing his light around looking for sharks. None of us saw any. But all of us heard the weird sound.

A swarm of the "rare," bright-orange Clarion angelfish. (Photo by author)

Finally out of air, we returned to the boat with our catch, which included dozens of the prized Clarion angels. Immediately, we began to discuss the puzzling noises we had heard. On a previous occasion, Bob and I had heard a mysterious moaning sound when night diving in the wreck of a fishing boat at the tip of the Cape. We mentioned it to a local Mexican fisherman, who said, "Eet ees the cry of the hongry tiger shark," but we dismissed that story as simply his attempt to scare us. In all our night diving around the world, none of us had ever heard anything quite like this sound. We were still baffled when we called it a night and hit the sack.

The next morning after a hearty breakfast we looked out from our cove and were astonished to see a humpback whale leap from the water and come crashing back down with a giant splash. All of a sudden the strange sounds made sense: they must have been the song of the male humpback! We were familiar with the recordings Roger Payne made in the late 1960s in the Caribbean, but what we'd heard underwater didn't sound anything like Payne's record. One striking feature that was missing was the echo of the whale song from the bottom of the deep ocean. Since we were in shallow water next to an island, we must have received the sound directly from the whale, without an echo. When submerged, too, our ears may not be very efficient at hearing sounds. In any case, we were delighted to have solved that particular mystery.

We later learned that the Revillagigedo Islands are the winter home for a large number of humpbacks, a little-known fact, mainly because of the islands' remoteness. The Hawaiian Islands, and in particular Maui, are of course famous as a wintering ground for humpbacks, where they come after feeding on the swarms of krill and herring in the seas around Alaska. Cetacean researchers have recently observed a few individual whales both in Hawaii and at the Revillagigedo Islands—an indication that at least some migration is taking place between the two wintering locations.

We saw a number of humpbacks during our days at Socorro, but because our primary objective was to collect fish we made only one unsuccessful attempt to swim with one. Still, it was exciting to know that while we dived, they too were diving in the waters nearby.

SOCORRO SHARKS

Now that our first night dive in the allegedly "shark-infested waters" was behind us, we felt much more comfortable about collecting, either day or night. Making steady progress, within a few days we had a good population in the holding tank on *El Navegante,* and we could now focus on other activities.

I had been taking pictures of some of the unusual fishes and invertebrates found on these islands. One day Chuck Nicklin and I were out in the inflatable Zodiac, and I rolled over the side into the water with my camera. Upon righting myself, through the cloud of bubbles I saw a small Galápagos shark coming straight at me at high speed. I raised my camera just in time to hit it on the nose. The commotion I'd made falling into the water must have excited the shark; we decided that it would be wiser for us to slip quietly in rather than with our usual backward roll and splash.

Chuck, who was shooting with an underwater sixteen-millimeter movie camera, wanted to get some shark footage, so we decided to

Something down there has sharp teeth! Kym Murphy having second thoughts about his next dive. (Photo courtesy Chuck Nicklin)

check out a wave-washed rock a quarter mile offshore from the main island. Known as O'Neill Rock, it had a reputation among the long-range sportfishing boats as being a hot spot for big game fish—and for sharks. In fact, the sharks were reportedly such a problem the fishermen would frequently bring in just the head of their prized tuna or jack.

We anchored the Zodiac next to the rock. Then, slipping quietly into the water with camera and bang sticks, we descended to a flat ledge at a depth of about forty feet. Within just moments, several six- to eight-foot silvertip sharks came up out of the depths to check us out. It was remarkable how quickly they sensed our presence. Silvertips are beautiful animals, with graceful bodies and very small almost silvery marks on the very tips of their dorsal and pectoral fins. These sharks, however, beautiful though they were, were getting a little too close for comfort, so we retreated to the safety of our Zodiac.

A seven-foot silvertip shark comes spiraling up from the depths to check us out. (Photo by author)

Chuck still wanted more film of the silvertips. Bob Kiwala grabbed a fishing rod and dropped the baited hook down to the ledge below. He immediately hooked a five-pound black jack (*Caranx lugubris*) and frantically started to reel it in. Chuck was leaning over the side of the Zodiac holding his movie camera underwater. A silvertip came spiraling rapidly up and grabbed the jack just before Bob got it to the boat. Sharks have very poor brakes, though, and, unable to stop, the shark banged into the side of the boat a split second after it chomped off the body of the black jack. Bob tried again, with exactly the same result: he ended up with just the head. Now we understood why the sportfishermen had so much trouble getting whole fish on board.

At that point we decided we'd better quit. It didn't seem like such a smart idea either to go back in the water with a bunch of excited silvertips or to have them slamming headfirst into our easily punctured rubber boat.

With the live-well on *El Navegante* almost full, we decided to take a little time off from collecting to explore the island. On the west side of Socorro there's a beautiful high rock archway, beyond which is a large lagoon. This lagoon had the richest growth of coral we'd seen on Socorro, with an abundance of happily grazing parrotfish.

True coral reefs don't occur on the western sides of continents but are found only on eastern shores and around islands throughout the central Pacific and Indian Oceans. The reason is simple: For healthy growth of coral to occur, water temperatures must remain relatively high and stable throughout the year. Yet along the western coasts of continents, upwelling brings cold water up from the depths at periodic intervals. This upwelling no doubt limits the profuse growth of coral on these shores and adjacent islands like Socorro.

The lagoon behind the rocky arch, however, was shallow, and because of the narrow entrance there was limited exchange of seawater with the outside ocean. It was therefore considerably warmer than the waters outside, which perhaps explains why the coral growth there was so abundant. The lagoon was ringed by a beautiful sandy beach, beyond which grew trees and brush. Everywhere we looked among the trees were bright red land crabs (*Gecarcinus planatus*).

Hidden among the trees we found a long-abandoned cot made of driftwood lashed together with strands of vine. Although it looked very old, there was no way to tell when it was built; everything ages so rapidly in the humid atmosphere of the tropics. Carved on one piece of the wood cot was the name "José." José may well have built this bed to keep himself out of the reach of the nocturnal land crabs. We wondered what his story might have been. Was he a shipwrecked fisherman whose boat sank, or perhaps a deserter from the lonely military garrison on the other end of the island? José's story remains a mystery.

Seeing this old cot brought to mind the account of the California Academy of Sciences' research schooner, the *Academy,* that hit the rocks of this island and went down in a storm in the 1920s. All on board survived, but they were marooned on Isla Socorro until they were rescued by the U.S. Navy months later. They survived by eating fish and parrots. After seeing how common the land crabs were, I'm sure they may have tried eating them too. However, this type of red land crab is reported to store in its tissues the toxic alkaloids from the plants it eats. Perhaps the color is a red flag that says, "Don't eat me."

Every day since we'd arrived, the volcano at the highest point of the island, Mt. Evermann (named after Barton W. Evermann, the former director of the California Academy of Sciences), had been actively releasing wisps of steam or smoke. From the boat, we judged the mountain to be about two thousand feet high. One day two of the younger, more adventurous—and maybe more foolhardy—members of the team, Kym Murphy and "Muggs" Carlton, decided to climb to the top and check it out. We dropped them ashore at six in the morning with some food and their canteens of water. The plan was to pick them up at the same place that afternoon. At the appointed time, however, there was no sign of them. Only after a while did we spot two tiny figures quite a way up the side of the island: they were still coming down.

Finally our two scratched-up, footsore climbers arrived at the shore and exhaustedly clambered aboard the Zodiac. Their little outing turned out to have been quite an ordeal. What from the boat looked like sparse, low shrubbery was in fact tall, dense brush, which Murphy and Carlton had had to force their way through both going up and coming down. Our estimate of the mountain's height was also a serious misjudgment: we found out later that the mountain is not 2,000 feet but actually 3,700 feet high. On reaching the volcano crater, they verified that the wisps we'd noticed were, in fact, hot steam. Surrounding the steaming vent were numerous land crabs that had ventured too close and been cooked by a sudden blast of steam. All in all, the duo's adventure sounded a bit much, and I was sure glad I hadn't gone with them.

HOMEWARD BOUND

The weather had been excellent at the start, which meant we were able to anchor *El Navegante* close in to the island and collect much of the time by diving right off the big boat. Unfortunately, this didn't last. Soon the wind picked up, to the point where the skipper didn't feel safe anchored so close to the rocky shore. Wisely, he moved the boat

Back in San Diego, the author transfers a healthy but safely anesthetized Panamic green moray. (Photo courtesy Chuck Nicklin)

to a safer anchorage, where three big tuna boats had also pulled in to wait out the storm.

Any productive collecting was now out of the question, so we relaxed, read the books we'd brought along, or fished for fun or supper. Catching a seven-foot-long blacktip shark off the stern created one flurry of interest, but boredom eventually set in.

At one point somebody—and my hazy memory points the finger at Kiwala—suggested that we do a scientific taste test of the several brands of tequila we'd bought in Cabo San Lucas. Seemed like a great idea at the time, but it sure didn't the next morning. Luckily, the storm hadn't let up, so at least we didn't have to go diving and chase darting, elusive fish while nursing hangovers. I don't remember now which tequila came out the winner. Perhaps the test needs to be repeated—but with a dedicated scribe to legibly record the results.

Eventually the storm passed, and after a couple more days of collecting we had the live-well on the boat full of fish. The time had come to head back. The trip north from Socorro to Cabo San Lucas, where we would refuel, is what we call uphill, which meant pounding our way right into the swells coming down from the Gulf of Alaska. It was a rough, twenty-five-hour haul, and we were miserable. Anything that wasn't bolted down went flying. I spent most of the daytime hours lying flat on my back on the deck of the main lounge.

We refueled at the Cape, and three and a half days later we pulled into San Diego Harbor to unload the fish and collecting gear. We had pulled off another successful collecting adventure, and went on to create the first major exhibit of the beautiful and unique fishes from the remote Revillagigedo Islands. The tropical reef tank at Sea World was a spectacular sight, highlighted by dozens of bright-orange Clarion angelfish and crosshatch triggerfish, and the smaller focus tanks in the marine aquarium were now filled out with morays, scorpionfish, and a wide variety of exotic invertebrates.

There's no question that the rapid success of Sea World was due primarily to the marine mammal shows. They were first and foremost what people came to see. Once in the park, however, many visitors were fascinated by the fish and invertebrate exhibits. Seeing this, and having seen the reactions of visitors to the many non-mammal exhibits at Steinhart Aquarium, I felt that a good aquarium without performing mammals, located in an area with potential visitors, could well succeed.

In the 1950s I'd made trips with the UCLA ichthyology class to central California to collect in the rich tide pools near San Simeon. Located on the slopes above that little town is Hearst Castle, the ostentatious mansion built in the 1920s by the newspaper magnate William Randolph Hearst. Although off the beaten track, the castle was visited by six hundred thousand people in 1971. Highway 1, known for its beautiful ocean views, runs along the coast from Morro Bay north to Big Sur and Monterey—and right past San Simeon. This location struck me as a perfect place for a small marine aquarium focusing on the abundant and diverse marine life of the area.

I contacted Roy Marquardt, who had financed the 1964 California Academy expedition to the Sea of Cortez and had always been a big supporter of Steinhart Aquarium, and told him about my idea. Agreeing that it was a good project, he contacted a San Luis Obispo–area resident he knew who owned land along the coast just half a mile south of San Simeon. This man, too, became interested in the possibility of an aquarium on his property.

I was very excited to have found a financial backer and a potential site for the aquarium; it was beginning to look as if this project would work. The location was good, the water quality was excellent for an open or semi-open seawater system, and I felt sure a small aquarium could be self-supporting. I returned to San Diego full of enthusiasm and drew up plans.

That was 1973, the year we were surprised by a totally unforeseen event that affected everyone's take on reality: the first serious gasoline shortage hit the United States, and it scared everyone, especially Californians, who are married to their cars. There were long lines and short tempers at the gas stations, many of which ran dry, and gas prices skyrocketed.

At the time, it looked as if there was going to be a permanent shortage of fuel that would drastically alter the lifestyle of each and every American. The fear was that the automobile would no longer be our ticket to freedom, able to take us wherever our hearts desired.

As a result of the gas crisis, the San Simeon Aquarium project was abandoned and I gave up my dream. By the time the gas crisis was behind us, my life had taken another course.

12

ROUNDABOUT TO STEINHART AQUARIUM

One day in 1974, out of the blue, I got a phone call at home from Dr. John McCosker, who had been appointed director of Steinhart Aquarium after the unexpected death of Earl Herald. Newly on the job, McCosker needed help completing the Fish Roundabout, a major exhibit that had been started shortly before Herald's death. Would I come on board to help see this project to completion, McCosker asked.

Ironically, after I left Steinhart to go to Sea World, Earl Herald had learned to scuba dive. Discovering that he loved it, he took up underwater photography. In spite of a history of heart problems, and against his doctor's orders, he went scuba diving at Cabo San Lucas with Jeffrey Meyer (later chairman of the board of trustees of the Academy). Near the end of the dive Herald apparently stopped breathing, and he was discovered lying motionless on the bottom. Back at the boat, all attempts to revive him failed. An autopsy done at Cabo San Lucas gave a likely diagnosis of heart failure.

Earl Herald had visited several of the many aquariums in Japan and was especially intrigued by a donut-shaped tank at Shima Marineland that completely surrounds the viewers. Thus inspired, Herald and his architects designed—and the GHC Meyer Family Foundation financed—a new exhibit at Steinhart to be called the Fish Roundabout

("roundabout" being an English term for carousel or merry-go-round). Construction had begun when Herald died.

John McCosker, a doctoral graduate from the Scripps Institution of Oceanography, was a first-class ichthyologist who would, in time, become an authority on the biology of great white sharks. Back in 1974, though, he had little experience with aquariums, and the prospect of completing this major new exhibit was daunting.

When he called with his invitation to come back to Steinhart, the timing couldn't have been better. Dissatisfaction at Sea World due to changes in management and the continuous postponement of a decision to build a dedicated shark exhibit had me feeling somewhat restless. Although Steinhart didn't have a large shark exhibit, it had a great diversity of animals from around the world, and I saw many opportunities to create new exhibits in addition to getting the Fish Roundabout up and running. I accepted McCosker's offer, and Betty, Eve, Amy, and I pulled up our roots once again for the move back to the San Francisco Bay area.

Steinhart paid our moving costs, but in some ways this move was more difficult for us. Our daughters were now teenagers and they'd made close friendships that were broken by my career decision. Selling a house in San Diego and looking for and buying one in the Bay Area was another hassle. The girls and I drove up to Marin County so they could be there for the beginning of the fall school semester. The three of us lived in a motel until our San Diego house sold and Betty could join us and we moved into our new home. Somehow we survived.

Despite these worries, I was delighted to be back at Steinhart again and to see many of the exhibits I'd helped set up nine years earlier. It wasn't long before I was in the swing of things and again making collecting dives in Monterey.

I also worked on designs for exhibits that would show visitors something they'd never seen before. An animal can be interesting because of its appearance—bright color, unusual anatomy, or effective camouflage, for example—or because of an intriguing behavior. Displaying an animal with an interesting appearance is easy: you simply put it in an exhibit and the visitors can immediately see what's special about it. Showing an interesting behavior, however, isn't as simple. Some remarkable behaviors happen only occasionally, like reproduction or feeding strategies; these can best be shown with a video placed adjacent to the live animal exhibit. And certain behaviors go totally unnoticed by us land animals simply because humans don't have a sensory system capable of detecting them. Consider, for example, electric fishes. Two unrelated families, the mormyrids from Africa and the gymnotids from South America, navigate, communicate, and find food and mates in their dark, turbid world wholly through complex electromagnetic impulses. They "see" their world with electricity, much in the same way bats and dolphins use sound and hearing to "see" theirs. Even more alien to us humans, many animals "see" their world by detecting chemicals with an olfactory sensory system far more acute than ours.

I found other sensory systems fascinating and wanted to share this phenomenon with aquarium visitors. I'd designed some electric fish exhibits at the Freshwater Aquarium at Sea World, so I set out to do something similar at Steinhart. Aided by a novice's knowledge of electronics, a couple of stainless steel electrodes, and a stereo amplifier from the local electronics supplier, I converted the electric pulses of the fish to stereo sound that we humans could hear.

This added a new dimension to an exhibit of black ghost knifefish (*Gnathonemus petersi*). As the fish moved about the tank "seeing" their environment and communicating with their electricity, the converted sound from these impulses moved through the speakers. Without the use of graphics, which often go unread, an interesting aspect of this fish's life now came to life for the visitor.

If a behavior of an animal occurs frequently enough—on command, so to speak—it's possible to schedule times when visitors can see them do their thing. One of the world's most remarkable fish is the archerfish (*Toxotes jaculator*), which shoots droplets of water at high velocity to knock insects out of the air or off overhanging vegetation. Although this behavior is well known, few people have had the opportunity to see them do it.

I took on the project of setting up an exhibit of archerfish at Steinhart. I'd kept archerfish previously in a conventional exhibit tank and noticed that sometimes during their regular feeding if a piece of food happened to stick to the wall of the tank just above the waterline they would shoot at it, knock it down, and eat it. I wondered if they could be "trained" to shoot at something out of the water. Naturally, I set about to find out.

I took a piece of vinyl plastic and hung it just above the water surface. At feeding time I tossed the food into the water directly in front of the plastic. The next step was to have some food stick to the plastic, so the archerfish would have to shoot it down. Selecting moist pieces of ground meat, I lobbed small morsels toward the vinyl sheet. Some stuck, and sure enough, the archers leapt out of the water to grab the food or, if they were out of easy leaping range, shot it down. Raising the piece of plastic higher solved the jumping problem, and they were forced to use their liquid bullets to get the food. It took only a few days for the archers to figure out the new feeding scheme; in no time they were consistently shooting at any food that stuck to the plastic.

Next I stuck a few pieces of food to the plastic sheet prior to feeding time and suspended the sheet above the water. No problem: they figured that out right away. I then affixed a bull's-eye to a piece of plastic, like an archery target—to give the visitors, not the fish, a clue of what the fish were supposed to shoot at. I now had the makings of a regular demonstration of archerfish shooting. It was put on two times a day, and the archers performed well every time.

An interesting example of their learning ability—and perhaps also of their laziness—occurred when two or three fish gave up shooting at the food altogether and instead waited in the back until another fish shot the food off the target. They would then rush in and grab it before the actual shooter could turn around. Seems like there's always a cheater in the crowd. The fish world really isn't so different from our own.

Having evolved as insect-eating specialists, archerfish are attracted to the slightest movement above the water. One time, aquarist Al Castro leaned over the tank four feet above the water, and an archerfish shot him in the eye with such accuracy and power that his contact lens popped clean out. Archerfish are attuned to the tiny motions made by their insect prey, and we figured that when Al blinked, the archer instinctively shot at the movement. They truly are remarkable animals.

I had learned a lot about the capture, transport, and husbandry of several species of sharks during the experimental work we did at Sea World. Although Steinhart lacked tanks of a size suitable for large sharks, it did have the roomy tank that Herald had built in the hopes of someday collecting and displaying a coelacanth (*Latimeria chalumnae*). The coelacanth, considered to be a living fossil, is related to the fishes that left the world of water and eventually evolved into land-dwelling vertebrates—including, of course, us. More about the coelacanth later, but suffice it now to say that Steinhart had a tank available to use if I could come up with an appropriate species of shark.

Herald's would-be coelacanth tank was presently occupied by a collection of trout and salmon. I thought that if I could find a small species of shark whose natural habitat was around reefs and so was adapted to living in tight spaces, we could create a much more interesting exhibit. No sharks in California fit those criteria other than leopard sharks, and we already had quite a few of those attractive animals in other exhibits. In the tropics, however, there are several shark species that live around coral reefs; I thought one of these would work well in an exhibit the size of the coelacanth tank.

Bruce Carlson at the Waikiki Aquarium had recently flown in a couple of small blacktip reef sharks (*Carcharhinus melanopterus*) from Canton Island, in the Phoenix group of the South Pacific. These sharks reach a maximum size of about six feet and at birth are only about fifteen inches long. They seemed like perfect candidates: small enough to be transported from their native waters at reasonable cost, but also, as members of the requiem shark family, always active and very "sharky" looking.

Canton Island has an interesting history. Under joint American and British control, it became the stopover point in 1940 for Pan American Airways trans-Pacific Clipper seaplanes on their way to the Orient and Australia. During World War II it was used as a military base. In 1946 it reverted to civilian use as a stopover for trans-Pacific flights, a function that eventually became unnecessary when long-range jet aircraft came into use. A NASA space tracking station was located there for a few years, and when that shut down the United States maintained its presence with a small civilian crew, in the event the island might be needed again as a military base. In 1983 the United States pulled out and the island was turned over to the Gilbertese, becoming Kanton Island, part of the nation of Kiribati.

In the seventies, the U.S. Air Force made weekly flights from Hickam Air Force Base in Hawaii to bring in food and supplies for the civilian crew. This was how the Waikiki Aquarium flew their blacktips back to Hawaii. It seemed like an ideal opportunity to collect and transport some blacktip reef sharks to San Francisco.

Thanks in part to the organizing help of Bruce and the staff of the Waikiki Aquarium, I arrived on Canton Island with my shark transport boxes, oxygen pumps, and, of course, dive gear. Stepping off the C-141 military transport jet, I was immediately struck by the intensity of the sun. I'd forgotten how intense the midday sun's radiation could be right on the equator. When I was handed several bottles of sunscreen and told to use it or else, I fully understood why.

The next day Waikiki Aquarium's Ralph Alexander and I went scouting for small blacktips on the shallow flats inside the atoll lagoon. Like other coral atolls, which are formed when a coral-ringed volcano sinks back below the sea, Canton is a ring of land enclosing a large lagoon. The lagoon has one entrance channel connecting it to the outside ocean. Inside the lagoon we found the little sharks cruising close to shore.

We stretched a fifty-foot-long nylon barrier net straight out from shore and waited until we saw a couple of sharks working their way along the lagoon's edge. Then, sneaking up behind them, we made a wild dash, hoping to panic them into the net, where we could grab them. More than half the time the alert, lightning-fast little sharks doubled back and shot between our legs to freedom. Twice, though, our technique worked, and after a couple of hours of exhausting work, skinned knees, and impending sunstroke, we had two beautiful little sharks in the boxes on the truck. Perhaps they were the retarded ones, but at that point we weren't picky. We had sharks! After two more exhausting days we had a total of six.

Some of us, needing a bit of a break, decided to do some diving and photographing on the outside of the island. Les Gunther, a trustee of the Academy who had come along to assist us, said he'd stay and keep trying for sharks. We wished him lots of luck and we went off diving.

After our dive we arrived back and were greeted by a surprisingly relaxed Les Gunther. Figuring he must have given up and retired with a cool drink to the verandah, we asked him how he'd done at shark chasing. Smugly he said there were eight more sharks in the pool. We suspected he was pulling our leg, but when we checked, sure enough, there they were cruising around.

We knew he was hiding something, and after much pressure Les admitted he'd asked Seone Tau, known on the island as "Tonga John," to help him catch some sharks. Loading the truck with the collecting equipment, Les and Tonga John, together with John's dog Blackie, drove out to the shallows where the blacktips were known to cruise. Les set the net out in the water, but then, looking back, he happened to see Blackie at the edge of the water, his foot on a shark pinned to the bottom. At that point Tonga John told Les that the one thing Blackie most loved was to catch sharks.

As if to demonstrate, Blackie raced off after another one. In a matter of moments the dog had it beached. Picking it up gently in his mouth, he dropped the shark at Les's feet.

Knowing Gunther's fondness for teasing, we found his story a bit hard to swallow, so the next day we decided to go shark collecting with Tonga John and Blackie. We were exhausted by our method of catching sharks and willingly enlisted Blackie's eager and skillful help.

Indeed, Blackie turned out to be truly amazing. We would scan the lagoon surface for tiny shark fins; as soon as one was spotted, Tonga John gave Blackie the word and off he'd dash, heading for the seaward side of the shark. The shark, meanwhile, seeing him coming, would frantically dash back and forth trying to make it past him to the safety of deep water. Like an expert sheepdog, Blackie skillfully herded the shark into shallower and shallower water. Rarely did a shark get past him. Instead they eventually ended up running aground in the inches-deep shallows, where Blackie would either pin them down with his foot or pick them up gently in his mouth.

A couple of sharks had slight teeth marks on them, but they weren't serious. One time a shark sought revenge and bit back, getting Blackie's

After chasing the blacktip shark down, Blackie the shark dog pins down his quarry and proudly waits for us to catch up to him. (Photo courtesy Foster Bam)

tongue. He was bleeding, but that didn't slow him down. He simply lived to chase sharks, and he'd quiver with excitement waiting to go after them.

We kept our sharks in a couple of small cement, rock-lined fishponds on the base, left over from the Pan Am days. Supplied with constantly running seawater, they made ideal holding pools until the time came to pack the sharks up for the weekly flight back to Hawaii. We had to keep Blackie away from the ponds so he wouldn't jump in and catch them again. He had a hard time understanding why those particular sharks were off limits but others were okay. That concept was just a little more than he could grasp.

The shark collecting accomplished, we headed for the south end of the lagoon, where a few months earlier Waikiki Aquarium's director Dr. Leighton Taylor and others had been swimming with a young twenty-four-foot whale shark (*Rincodon typus*) they named "Mini." Apparently she'd come into the lagoon through the one entrance channel and wasn't able to find her way back out. People at the base were concerned that there wasn't enough food in the lagoon for the plankton-feeding shark. There was much talk among the island crew about how to get her out again to the open sea. One proposal even involved dynamiting a new channel through the coral reef.

We didn't see Mini. Instead we found another whale shark, a smaller male about thirteen feet long that had been given the nickname "Mickey." This was my second experience swimming with a whale shark, the first having been with a very large thirty- to forty-foot shark near Loreto in the Sea of Cortez.

In the shallow water of the lagoon the shark swam slowly, unperturbed by our presence next to it. The water at the end of the lagoon was rather murky, so our underwater pictures weren't the greatest, but it was still thrilling to be in the water with such a wonderful and docile animal.

The time came for the weekly Air Force C-141 plane to arrive with supplies and to fly our sharks and us back to Hawaii. For the sharks, I'd brought two large fiberglass transport boxes—big enough for eight little sharks (selected from the fourteen) to swim constantly—as well as oxygen, twelve-volt car batteries, and submersible oxygen pumps. The flight to Hawaii was uneventful, and the sharks arrived in perfect shape.

Waikiki Aquarium had arranged for a large pool that I could use to hold the sharks for three days until my flight to the mainland. This gave the sharks time to rest and recover in case there had been any physiological stress from the first leg of the trip. The next phase would involve two legs: Honolulu to Los Angeles, where I would have a six-hour layover to change planes for the flight up to San Francisco.

The sharks and tons of other freight were loaded in the Pan Am DC-8 cargo plane, and we took off from Honolulu just as the sun was setting. The crew consisted of two pilots, a navigator—and me and my sharks. After making sure the life-support system was functioning, I moved up to the cockpit as the heavily loaded plane roared down the runway, lifted off, and circled over Diamond Head. What a breathtaking sight, with the glowing red sunset and twinkling lights of Honolulu! As we sped eastward, the dark of night came quickly. It seemed unreal to be flying through space in this huge plane with boxes of live sharks—as if we were in another world disconnected from the earth below.

After a couple of hours the navigator checked with Los Angeles International Airport traffic control and was told that they were heavily fogged in, the airport was closed, and we were being diverted to Las Vegas. My heart sank. What was I going to do in Las Vegas with live sharks and no possibility of making seawater changes? The likelihood of finding another cargo plane to fly from Las Vegas to San Francisco was somewhere between slim and nil. I expressed my concern to the chief pilot, who fully understood my predicament. Not to worry, he said.

Apparently the flight schedules of cargo planes are pretty loose, and he called Pan Am's dispatcher, explained the situation, and asked if he could divert to San Francisco instead. He was told it was clear there and to go ahead. What a relief! Having L.A. fogged in turned out to be a blessing in disguise. Not only did the sharks and I end up right where we needed to be, but we also bypassed the long layover in L.A. plus the second flight north to San Francisco. It was an unexpected happy ending to an already spectacular flight.

At Steinhart, the blacktip reef sharks did well and grew at a rapid rate. After two years I became concerned that they would soon be too large for their 6,500-gallon exhibit. Since Steinhart didn't have a larger tank they could be moved to, we decided that it would be wise to find a new home for them before they became too large for our transport tank.

I called Sea World, knowing that they had two large heated shark systems our sharks could go into. The smaller of the two would be perfect until the blacktips grew large enough to be safe from being eaten by the eight-foot bull shark and ten-foot lemon shark in the main shark exhibit. The Sea World staff was more than happy to give them a home, and we went ahead with plans for the move.

Prior to moving any fish over long distances, it's standard practice to fast them for a few days. This gives them time to metabolize any food they've eaten and to rid their system of wastes that may contaminate the relatively small amount of water used in shipping, especially nitrogen waste in the form of ammonia, which is highly toxic to fishes. So we withheld food from our sharks for three days, then loaded them into the oxygenated transport tank on the truck and took off for San Diego. All went well, and the blacktips arrived at their new home in good shape.

BACK TO CANTON ISLAND

A second trip to Canton Island was planned to collect more small blacktips. This time I would go with Ray Keyes, the curator of fishes at Sea World, and Bob Kiwala. Ray also wanted some of the little blacktip reef sharks, and Bob needed to collect tropical sponges and tunicates for Dr. John Faulkner of Scripps. These defenseless-looking invertebrates, which manufacture toxic chemicals to keep other organisms from crowding them out, had recently become a hot item in the pharmacology industry's search for new antibiotics and possible cancer cures.

This time I was planning to use a different and less expensive method of shipping my sharks back to San Francisco and Ray Keyes's sharks to San Diego. Instead of large boxes, we would use four-foot-long Styrofoam boxes made for shipping long-stemmed roses, fitted with the polyethylene bags used by coroners for human bodies, which happened to be the perfect size. With a few inches of water, each oxygen-filled box would nicely hold two little blacktips snug in their body bags. These smaller containers could then go in the cargo hold of a passenger plane.

Arriving at Canton, we looked up Tonga John and Blackie, and they were willing—and in the case of Blackie, eager—to help us again. With Blackie's expert help, the shark collecting was completed in a couple of days.

The remainder of the time was spent diving on the steep drop-off on the outside of the atoll. I'd brought my underwater camera, so I spent some time taking pictures, and Bob still had to collect his sponges and tunicates for Dr. Faulkner and needed help with that.

The outside of the reef was the hangout for the gray reef sharks (*Carcharhinus amblyrhynchos*). These sharks, which can be territorial, are known for the warning display they put on when their personal space has been invaded. They arch their backs, lower their pectorals, and go

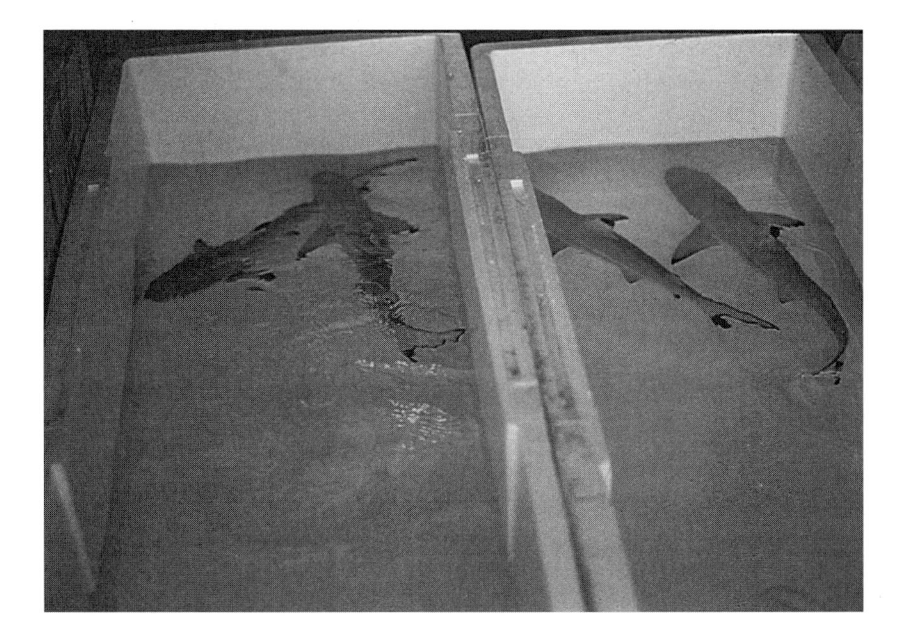

Blacktips resting in shipping boxes. (Photo courtesy Ray Keyes)

through slow, highly exaggerated swimming motions. If the intruder doesn't withdraw, they attack at lightning speed.

Bob was naturally concerned that with his attention focused on looking for tunicates on the reef in front of him, he wouldn't be able to keep an eye out for gray reef sharks that might become upset by his presence. So I went with him and, riding shotgun with my shark defense bang stick, kept an eye out for possibly ill-tempered gray reefs. Plenty of them came by to check us out, but they just looked and kept on going.

It was a real pleasure to have time to enjoy the incredible diving on this pristine coral atoll. There were many memorable sights. A major one was the giant manta rays with their twelve-foot wingspans that came gliding by to check us out. One could only stop and stare in wonder at the beauty of a passing school of electric-blue-and-yellow fusiliers or the occasional glimpse of a lone majestic Napoleon wrasse.

Having spent most of my diving career in the cold, sometimes murky waters of California, I found it a delight to dive in eighty-five-degree water, with no need for a cumbersome wet suit and heavy weight belt, and the hundred-foot-plus visibility was utterly enticing—so much so, in fact, that we really had to watch ourselves. It was very tempting to keep going just a little bit deeper, to see what wonderful new animal we might find. The usual clues of getting colder or darker with increasing depth simply weren't obvious, so it was important to keep an eye on our depth gauges. The nearest decompression chamber was two thousand miles away in Hawaii, and with only one flight a week, a case of the bends would be big trouble.

In addition to collecting sharks, we made several night dives to collect a variety of coral reef fishes, including some of the beautiful South Pacific parrotfish, which were destined for Sea World and the Scripps Aquarium. We kept them in one of the concrete pools similar to the one we used for the blacktips. At one point Bob commented that the parrotfish sure were hard to see in that pool. I told him not to worry; they must just be hiding under the rocks.

However, when the time came to pack up our sharks and fish for the flight to Hawaii, we couldn't find a single parrotfish. We were baffled. We knew we'd put them in there. We knew they couldn't get out; yet we found no dead ones. Where'd they go? One of the island crew finally took pity on us and explained. It seems that the night operator of the nearby desalination plant was Samoan, and one of the delicacies in Samoa is raw parrotfish liver. Every evening after we went to bed he would catch one of our beautiful parrotfish, eat the liver for his dinner, and toss the rest in the ocean. A little lesson in cultural differences.

The week drew to an end and it was time for us and our precious little sharks to catch the weekly flight to Hawaii and on to the mainland. The trip back was a far cry from the flight two years earlier on the Pan Am freighter. This time the sharks were in the cargo hold, and to our fellow passengers we just looked like well-tanned tourists returning from a Hawaiian holiday. The blacktips did well, and again, after two years at Steinhart, they grew too large for their exhibit tank and ended up at Sea World, along with the ones that Ray Keyes—or rather Blackie—had collected.

There was a rewarding sequel to this second trip to Canton. At Sea World in 1996, an adult blacktip reef shark gave birth to a litter of pups. They were the same size as those we collected in 1976. It was really good to hear that a male and a female from the original sharks that Blackie, the shark-catching dog, had helped us collect had grown to maturity and, twenty years later, had reproduced. Successful reproduction of elasmobranchs in aquariums has been rare, but it's becoming more and more common as public aquariums build more spacious, more naturalistic environments for them. I hope the days of keeping large sharks in small tanks are over. Responsible aquariums no longer keep sharks they know will grow too big for their facility.

THE GREAT WHITE SHARK EXPERIENCE

Although the daily activities of an aquarist may seem to be fairly predictable, there are times when a single phone call causes you to drop whatever you're doing and instantly redirect your energies. One day I received such a phone call from Pete Halley, descendant of the discoverer of the comet.

Pete and his brother were commercial halibut gillnetters working out of the little town of Marshall on Tomales Bay, north of San Francisco. They'd set their net over a sandy bottom outside the entrance to the bay and, when pulling it to check their catch, found a young great white shark that was still very much alive. Through the marine operator they called Steinhart Aquarium and asked if we were interested in the shark, and if so, what they should do to keep it alive.

They had no live tank on board, so I suggested that they somehow get a long line on the shark and tow it slowly back into Tomales Bay, where we'd meet them at their dock. Aquarist Ed Miller and I dropped what we were doing and loaded the collecting truck with the largest fiberglass holding tank we had, six feet by three feet, an oxygen cylinder, regulator, oxygen pump, shark stretcher, and hip boots, and we took off for Marshall.

Three hours later we were met in Marshall by the rather glum-looking Halley brothers. They said that they'd managed to get a line on the shark and all went well for a while with the shark swimming slowly behind the boat on its long leash. At one point, however, the shark swam under the boat and the line became wrapped in the boat's propeller. Of course, the boat came to an immediate halt. It took a little while to get the tightly wrapped line free from the prop, during which time the shark was no longer able to swim. By the time they docked, the shark wasn't doing well and it sank to the bottom.

Ed and I walked out to the end of the wooden pier and, looking down through the shallow water, saw the shark lying on its side on the muddy bottom. Showing no gill movement or other signs of life, it looked like a dead shark.

I didn't tell the Halleys I thought it was hopeless, though. They had made such a valiant effort to bring the shark in alive, and we'd driven all the way up there, so I decided at least to go through the motions to try and revive it. After starting up the oxygen pump in the transport tank, we towed the shark alongside the pier to the shore. Ed and I waded out and got it into the shark stretcher, then the four of us carried it ashore. Hoisting it up into the truck, we found that it was a foot longer than our six-foot tank and its tail stuck out of the top.

The shark had been just a few minutes in the highly oxygenated water when, surprisingly, its gills began to move. We had given it up for dead, but it was now breathing regularly and moving its tail. We couldn't give up on it now, so we took off for San Francisco and the Steinhart, stopping a couple of times along the way to check on it. Each time we found it was still breathing well and its muscles were going through swimming motions despite its cramped quarters.

Back at the aquarium—which luckily was now closed for the day, so we didn't have to deal with a crowd of curious visitors—four of us lifted it out of the tank into a shark stretcher, and the electric hoist took it up to the top of the Fish Roundabout. Lowered into the exhibit, the shark started slowly swimming away, but after a few strokes of the tail it stopped and sank to the bottom.

John McCosker was already suited up, and he jumped in to pick the shark up to get it going again. It worked: the shark's tail strokes began one more time, but with the same result. After a few beats it stopped and sank to the bottom. Obviously the shark wasn't dead, since it could swim, so we weren't about to give up on it. We took turns "jump starting" the shark all night, but by morning its self-swimming periods were becoming shorter and shorter; we finally admitted defeat and pulled it out of the tank.

John McCosker helping the great white shark in Steinhart's Roundabout. (Photo courtesy Al Giddings Images)

One observation I made during that long night was that the "look" of the shark just wasn't right. Although hard to describe, something was missing. Its eyes had the blank look of someone in a trance, with their eyes open but not really seeing. Since that day I've seen that look in the eyes of other sharks and have concluded that the brain behind those eyes was no longer functioning. The nerves and muscles involved in swimming and breathing continued to work, at least to some degree, but the animal was swimming dead. At that point there was probably no way to repair the damage done and bring the shark back to life. In spite of our wishes for its recovery, we were wasting our time going through those cold, numbing motions again and again.

White sharks are considered "ram" ventilators: they need to swim in order to force oxygenated water across their gills. A period of nonswimming can result in a critical lack of oxygen to the brain, leading to damage and the eventual death of that organ. Without a fully functioning central nervous system, these animals can never survive. Precisely how long an immobilized shark can survive without oxygen is unknown. No doubt it varies with each species, and many other variables have an effect as well.

I later had the opportunity of swimming with a fully alert great white shark in Monterey. Its appearance and the way it looked at me were dramatically different. I could tell immediately that there was a living, responding being behind those jet-black eyes.

13

SEARCH FOR A LIVING FOSSIL

ALTHOUGH I DIDN'T KNOW it when I took the job of curator, my arrival at Steinhart couldn't have happened at a better time. John McCosker had organized an expedition to the Indian Ocean to try and capture a live coelacanth—the fulfillment, if successful, of Earl Herald's dream. And of course, I would be going along.

The legendary coelacanth, known as a living fossil, has intrigued biologists the world over. The entire group, many examples of which exist in the fossil record, was believed to have died out seventy million years ago. In 1938, however, a live, five-foot-long coelacanth was caught off the east coast of South Africa by a trawler. Since then a number of them have been captured, almost all around the Comoro Islands north of the island of Madagascar. The Comoros were believed to be the sole location of the remaining population of coelacanths, but surprisingly, in 1998 two of these remarkable, ancient fish were caught near Sulawesi, Indonesia, one of which was even kept alive for a short while.

The living coelacanth is named *Latimeria,* after Ms. Courtney Latimer of the East London Museum. Seeing the fish on the trawler, she noticed a resemblance to the fossil fishes she had studied in her paleontology books. The fish, which appears to have changed little in three hundred million years, possesses characteristic lobed fins with bony leg-

Coelacanth. (Illustration by Charles Douglas. Reprinted by permission, Canadian Museum of Nature.)

like supports, similar to those that evolved into the four limbs of land vertebrates. They are large fish, reaching a length of six feet or longer and weighing up to 220 pounds. No living coelacanth had ever been kept in an aquarium. If our expedition were successful, not only would it be the first of these fascinating creatures ever to be on display, but scientists would flock to the Academy to study it.

JOURNEY TO THE INDIAN OCEAN

The expedition was planned for February 1975 to coincide with the month the monsoon usually arrives and also the month when most coelacanths have been caught incidentally by local Comoran fishermen. The actual target of these fishermen is the oilfish (*Ruvettus* spp.), which is used for food. Oilfish are caught with handmade baited hand lines from small pirogues, or dugout canoes, in the deep water close to the islands.

An earlier and unfortunately unsuccessful coelacanth expedition in 1971 was led by Dr. Murray Newman, director of the Vancouver Aquarium. To stimulate the fishing effort, he offered a handsome reward to any fisherman who caught a healthy coelacanth. The lucky fisherman would receive a free, all-expenses-paid trip to Mecca plus a cash bonus. This was strong incentive indeed, since Comorans are devout Moslems and none of the fishermen could ever expect to afford such a trip in their lifetime. John McCosker followed Murray's example and made a similar offer.

The expedition included John McCosker; Sandra McCosker, anthropologist and French interpreter; Dr. Michael Lagios, physiologist; Dr. Sylvia Earle, specialist in marine algae and a superb diver; Al Giddings and Chuck Nicklin, underwater filmmakers; John Breeden, son of the financial backer of the expedition; and me, who, we all hoped, would handle the transport of the first living coelacanth back to the Steinhart.

What with all the diving, photographic, and fish transport gear, we had literally tons of equipment. It was quite a project just getting it from San Francisco to that remote corner of the Indian Ocean. Getting the team there wasn't simple either, mostly because we didn't all plan to arrive at the Comoros at the same time. Sylvia Earle and I, for example, came later than the others, and on separate days.

My itinerary was San Francisco to Houston to Paris, where I would catch the twice-weekly flight to Dar es Salaam, Tanzania, and from there by Air Comores to Moroni on Grande Comore. The schedule looked fine on paper, but my flight from Houston to Paris was delayed, so I missed the connection to Tanzania and had to spend three days in cold, rainy Paris. Since I'd packed clothes for summer in the Southern Hemisphere, I froze.

Finally, though, I was on board the Air France flight from Paris to Dar es Salaam, and it was spectacular. Not only was the French food delicious, but the accommodating pilot circled and banked the plane twice around the top of Africa's highest mountain, nineteen-thousand-foot Kilimanjaro, giving passengers on both sides a breathtaking view of its majestic, snow-covered peak.

Upon my arrival in hot, dusty Dar es Salaam, the treatment I got from the Tanzanian immigration and customs officials was far from welcoming. I was separated from the rest of the passengers and secluded in a small, bare room with a silent teenage guard wielding a loaded submachine gun. Four hours later, without explanation, they released me to board the Air Comores DC-3 for the two-hour flight over the Indian Ocean to Moroni, the country's capital.

As we flew in low, the island, with its lush, tropical vegetation, was a beautiful sight. Edging the sea was black, volcanic rock with a narrow band of azure blue shallows that dropped precipitously into the dark blue of deep water close to shore. Just seeing it made me anxious to dive and see firsthand the exotic Indian Ocean creatures I'd seen only in books.

John McCosker had arranged for our group to stay at the appropriately named Hotel Le Coelacanthe, run by French proprietress Cici. Immediately beyond the hotel was a flat expanse of lava; at its edge the ocean dropped down to a thirty-foot-deep shelf, and from there plummeted steeply to several thousand feet.

Sylvia and Al were the most experienced deep divers among us, and while we were there they made a number of dives below two hundred feet. They left extra scuba tanks on the thirty-foot shelf to provide air for their decompression stops so they could safely return to the surface. Sylvia was so comfortable and relaxed underwater that she often had air left in her first tank even after their decompression.

LES PETITS PEUGEOTS

French divers on the island told us about a little fish with bright lights that they saw only at night. They'd given it the nickname "Petit Peugeot" because of its two little "headlights" that blinked on and off. One night we made a dive during the dark of the moon in front of the hotel to see if we could find them. Dropping down to eighty to a hundred feet, we stopped and turned off our dive lights. I'd seen bioluminescent sea pens and plankton in California, but I was not prepared for the brilliant and continuous flashes of light these flashlight fish (*Photoblepharon steinitzi*) produced. They seemed to be everywhere, and the deeper we went, the more there were.

Although the coelacanth was still our primary quarry, we decided to focus some of our efforts on collecting this second fascinating creature.

A flashlight fish showing its bright light organ below the eye is out cruising in the dark of night. (Photo by author)

We optimistically set up a holding tank filled with seawater in a darkened storeroom in preparation for the Petits Peugeots—assuming we could figure out how to catch them. Then we made another night dive in the same area. This time I took my hand net, plastic bag, dive light, and some small-gauge hypodermic needles to deflate their swimbladders. Reaching the depth where we'd seen them earlier, we turned our lights off. Just as before, there they were all around us, happily blinking away.

I swam slowly toward one with my light out, hoping I wouldn't run smack into a rock. I clicked the light on and aimed it right at the fish's eyes. The fish stopped, and with a quick swish of the net I had it. When transferred to the plastic bag it swam round and round rapidly blinking its lights. Not all the flashlight fish were as easy to collect as that first one. We learned that swimming toward one and turning the dive light on too soon was a sure way to spook it. Once alerted, it would dart off at high speed and be gone.

After a couple more dives we had a half-dozen flashing fish swimming around in our darkened holding tank. We learned that although there were many more fish down deeper, below a hundred feet, their survival wasn't nearly as good as with fish caught in shallower waters. Even though we deflated their swimbladders with the needle, they seemed to suffer from the bends, much as a human diver would coming from the same depth.

One of the French divers told us that the best place to see the Petits Peugeots was just off the end of the runway of the new airport. Borrowing a couple of jeeps, John McCosker, Sylvia Earle, John Breeden, and I headed out there after dark one moonless night. A very steep, rocky dirt road led from the runway down to the water's edge. After getting our dive gear on—which wasn't easy in the pitch black—we slipped into the water and headed down.

Our buddy system fell apart in the dark, and I found myself alone after descending to fifty feet. There I saw only a couple of flashlight fish, both exceptionally wary and unapproachable. Not wanting to go down to a hundred feet, where there were probably more fish but where their survival would be poor if I caught any, I started up into shallower water and ran into Sylvia, who was also wandering around alone.

Not quite knowing where we were or where we were going, we somehow ended up at the mouth of what appeared to be a cave. Entering the cave, we found that it kept going for quite a way and also grew shallower in depth. Soon we saw sparkling flashes of light ahead. Getting closer, we realized that dozens of blinking flashlight fish were clustered in front of us. It was an amazing sight.

I looked at my depth gauge and saw that we were in about ten feet of water but with a solid rock roof over us. It dawned on me that the only way to reach the surface was to go back the way we had come in: there was no going straight up. We had plenty of air, so there really wasn't anything to worry about. Meanwhile, in front of us was a veritable diamond mine of trapped flashlight fish. Instead of seeking refuge from daylight by going deep, these fish had found darkness in this shallow, underwater tunnel. Unhitching my hand net, I started netting and bagging as fast as I could. It didn't take long before we had as many fish as I felt would be safe to carry in two plastic bags.

We headed back out of the cave, which we realized later was a submerged lava tube. Lava tubes are formed when the surface of flowing lava cools to form a tube and still-molten lava continues to flow inside the solid tube; when the supply of lava stops, it drains out, leaving the tube hollow. We made our way back to where we'd entered the water an hour earlier. Climbing out on the rocky shore, however, we found we were alone. One jeep appeared to be stuck at the bottom of the steep hill, and the other jeep was gone. We figured the others had been diving deep and had used up their air long before we did. Presumably they'd gone to get help for the stuck jeep and would return shortly.

Sylvia and I hiked up the dirt road to the end of the new airport runway. Patiently waiting for our ride, we lay on our backs on the deserted tarmac and gazed up at the galaxy's millions of bright stars, our two buckets of brightly blinking fish between us. What a surreal experience this was. Eventually the others came back and pulled the stuck jeep free, and we all piled in and headed back to the hotel.

Excited about our catch, we banged on Chuck Nicklin and Al Giddings's door and woke them up to show them our glowing buckets, our prize catch. Chuck took one look and said, "Big deal! We saw thousands of those this evening," then rolled over and went back to sleep.

And it was true. In addition to all their video and still camera equipment, Al and Chuck had brought an underwater electric-powered scooter called a Farallon. Renting a powerboat at the harbor to take them and their gear out, the two of them made a deep night dive along the drop-off. With Chuck hanging on to Al's fins, they cruised effortlessly along the face of the drop-off at a depth of two hundred and fifty feet. They were hoping to see and film a coelacanth that might have come up from the depths during the night. They saw no coelacanth, but they were surrounded in that black sea by a thousand blinking flashlight fish.

During the day we had a good chance to see how the flashlight fish is put together and how it can instantly turn its bright lights on and off. Just beneath each eye is a large white patch containing a dense colony of brightly bioluminescent bacteria. Below that is an eyelidlike structure that the fish can quickly flip up to cover the light organ. The bacteria produce light continuously, but the fish effectively turns the light off by covering it with its shutter.

It is believed that the fish uses this light organ for at least three purposes. One is to attract food, in the form of zooplankton that are drawn to the light; the second is to communicate with other flashlight fish;

and the third is to confuse predators, since the blinking makes it very difficult to pinpoint a moving flashlight fish's location. (This last strategy certainly worked on us!)

With a newspaper, we tested the brightness of their light organ. Two recently dead flashlight fish, we found, provided quite enough light to read the paper easily in a dark room.

The flashlight fish turned out to be one of the highlights of the trip, but we saw many other wonderful creatures as well. One particularly fascinating creature was the three-inch scorpionfish called a leaffish (*Taenianotus triacanthus*). Leaffish live in shallow water in areas with profuse algal growth, and they come in a variety of colors—red, orange, tan, brown: much like shades of seaweeds. I caught two leaffish and when I placed them in the bottom of a bucket they immediately started rocking back and forth as though they were pieces of red or brown algae swaying in the surge. Obviously unaware that their normally effective rocking behavior was a total failure in a bare bucket, the two fish weren't even moving in unison. We found their out-of-place protective behavior rather comical, and I released them unharmed where I'd picked them up.

The variety of fishes and invertebrates we saw was mind-boggling. On every dive I came across species I'd never seen before nor even read about in books. I think I could have spent a year there before I'd be able to say I was familiar with most of them. All this newness certainly made each and every dive exciting.

On one dive Sylvia found a giant red and orange nudibranch (*Hexabranchus sanguineus*) that weighed several pounds. She named it "Flower" and brought it up to show us, then returned it to where she'd found it. We were most impressed with her colorful pet! We saw morays of several species and sizes, including a beautiful pair of honeycomb morays (*Gymnothorax favigineus*) soliciting cleaning services from a blue-and-white cleaner wrasse (*Labroides dimidiatus*).

I was already familiar with the striking moorish idol and imperial angels (*Pomacanthus imperator*), but I found the schools of little fairy basslets charming (*Anthias* and *Pseudanthias* spp.) as they flitted among the corals. These I'd never seen in an aquarium, and two or three species of them seemed to be everywhere. Basslets are one of those fishes, like

A pair of honeycomb morays soliciting the services of the little blue-and-white cleaner wrasse. (Photo by author)

many of the wrasse family, that have a harem arrangement, with a dominant male and a number of females. If something should happen to the male, the dominant female in the group changes sex—as well as size and appearance, often quite dramatically—and becomes the male of the harem. We humans find that a bit kinky, but it's a fairly common reproductive strategy in the world of fish.

One type of animal was conspicuously absent: on all the dives we made we didn't see a single shark. Perhaps the Comoran pirogue fishermen had fished them all out over the years. Not that we really missed them; it was a pleasure not having to think about what might be coming up behind you.

In sharp contrast to the beauty we saw underwater were the living conditions of the Comoran people. Most of the arable land was owned by the French and used for growing vanilla beans for export to France, an industry that did little to benefit the Comorans. When we were there the islands were on the verge of gaining independence, but they would not be left with much when the French left. In fact, after independence was achieved in 1975, the Comoros were listed by the United Nations as the poorest country in the world, with virtually no health care or sanitation systems and little in the way of schools.

Shortly after we arrived we heard that a number of people in a village on the other side of the island had died of what was suspected to be cholera. Each day we heard of more deaths, occurring a day and a half after the first signs of illness appeared, and then we heard of people dying in other villages. There were no doctors for the Comoran people, and the French seemed not to care. They just drank their Perrier bottled water and ignored what was going on. After fifteen days more than two hundred people had died. The Comoran president's solution was to leave the islands so he wouldn't get sick himself. We found his callous attitude very hard to understand.

THE ELUSIVE COELACANTH

The days and nights passed, but still there was no sign of the monsoon that would bring an increased chance of catching a coelacanth. We had a deadline for when we had to leave, with no possibility of staying longer. So John McCosker arranged with the director of Le Coelacanthe Laboratoire in Moroni to obtain two frozen coelacanths that had been caught the previous year.

During our final few days on the island we started packing the mountain of equipment we'd brought and made plans to get the forty precious flashlight fish back to San Francisco. Finally the last day of our stay arrived. We packed the live fish in bags and Styrofoam shipping boxes and the two big frozen coelacanths with dry ice in the fiberglass tank we'd used to hold the flashlight fish. Then we boarded Air Comores for Dar es Salaam, to connect with Air France for Paris.

Our arrival in Paris was chaotic, what with the gear and especially the highly perishable live and frozen fishes. Because we had no live coelacanth to care for—my appointed task—I took the opportunity to fly to London to visit my sister, Marcia, who lived in England. John McCosker had the job of shepherding the flashlight fish to New York and on to San Francisco.

Somehow he made it, along with half the fish, which survived in spite of being loaded in a non-temperature-controlled cargo hold. Their water temperature had dropped from eighty-five degrees down to the forties; it was amazing any made it at all. On arrival in San Francisco the fish were whisked away in the Steinhart truck, but an exhausted John had to conduct a press conference. Hearing about this later, I felt rather guilty sticking John with the task of enduring it alone.

The flashlight fish were exhibited in a completely dark 1,300-gallon tank. It was fascinating to watch them blink their lights on and off as they swam around among the rocks. Indeed, we could easily see how effective the flashing lights would be in confusing predators, since we found tracking a single fish as it darted about quite impossible.

The two frozen coelacanths made it in perfect shape. One was preserved in formalin and then alcohol and put on exhibit at the Academy, and the other went to Scripps. Although we hadn't managed to collect a live coelacanth, the expedition was still a success: it resulted in the first exhibit in the United States of bioluminescent fishes as well as publication of the most complete scientific papers to date on coelacanth biology.

Yet even with these publications, the fact remained that almost nothing was known about the natural history and behavior of these remarkable living fossils. This gap in our knowledge was narrowed beginning in 1987, when Hans Fricke of the Max Planck Institute in Germany made a series of submersible dives at the Comoros and filmed a number of live coelacanths he discovered living in deep submarine caves.

Since our expedition in 1975 there has been one other serious attempt to collect a live coelacanth. In 1989 Toba Aquarium of Japan launched a multimillion-dollar effort to capture two live coelacanths to place on display, hoping, somewhat unrealistically, that the pair would be a male and a female that would then breed. They used a research vessel deploying a submersible ROV (remotely operated vehicle) equipped with specially designed traps. That effort also was unsuccessful.

Because of the difficult logistics and expense of a collecting trip and the deepwater habitat of the coelacanth, plus recent efforts to list this unique fish as an endangered species, it's an open question whether a coelacanth will ever be kept in an aquarium. However, in September 1997 and July 1998 two coelacanths were captured off Indonesia; this totally new location means the species is more widely distributed than previously thought. If other populations are discovered, they may offer an opportunity for an aquarium or research institution to collect and possibly keep a live one. Time will tell.

The Comorans are practical people, and they eat coelacanths when caught. Yet they don't specifically fish for them because they aren't that tasty. One evening, Cici at Hotel Le Coelacanthe prepared us some coelacanth she'd stashed away in her freezer. We found it to be excellent, though I suspect it was Cici's French culinary skills that made it so good.

Shortly after we got back to San Francisco, SPOOF (Society for the Protection of Old Fishes) held its annual meeting, and instead of rubber chicken for the group banquet we dined on coelacanth, sturgeon, bowfin, and gar, all "old," primitive fishes. The coelacanth wasn't nearly as good as Cici's.

14

TO CHILE, EASTER ISLAND, AND RAROTONGA

ONE DAY IN 1979 Dr. Alfredo Cea Egaño, from the University of North Chile, visited John McCosker at the Steinhart to discuss a project in the coastal city of Coquimbo. Quite persuasively, he talked about the university's plans to build a marine biology station and public aquarium. John was so impressed with Dr. Cea's proposal that he offered the cooperation of the Academy and volunteered my services to go to Chile and help Dr. Cea design the aquarium. I would receive my salary from the Academy, while the University of Chile would pick up my travel expenses.

It was a wonderful opportunity, and fortunately I wasn't in the middle of any projects that couldn't wait until I got back. Still, although I looked forward to it, I was a little apprehensive. Just a few years earlier, a military coup had overthrown elected president Salvador Allende—a coup in which the U.S. Central Intelligence Agency was implicated. Now right-wing dictator General Augusto Pinochet was in command, and there were disturbing reports of assassinations, disappearances, and other human rights violations. Chile, I concluded, was the kind of place where it would definitely be prudent to keep your mouth shut and your thoughts to yourself.

On the plus side, I hoped I would have an opportunity to dive with Dr. Cea, who himself was an avid scuba diver, and see some of the rich marine life along the coast of Chile. The Mediterranean climate of the coast is similar to that of California and South Africa. Both have a cool current that flows from the Poles as well as upwelling of nutrient-rich water. This upwelling supports abundant plankton, which in turn feeds huge schools of anchoveta—a small fish that forms the basis of the world's highest-tonnage fishery, a critical industry in Chile. And of course this productivity continues up the food chain, for anchovetas feed flocks of cormorants, which then produce nitrogen-rich guano deposits, found on the offshore islands of Chile—another important natural resource that in the past contributed to the country's economy.

ARRIVING IN SANTIAGO

I was met at the airport in Santiago, the capital of Chile, by a friend of Dr. Cea. He'd arranged for a hotel in town for my overnight stay and provided directions on where to catch the bus to Coquimbo the next morning. Once I'd settled in at the hotel I had a couple of hours to look around the city. Like other cities in the world, Santiago had shops, bustling crowds of people, and too much traffic. The striking difference here was that standing on most street corners were soldiers carrying submachine guns. This was definitely not America.

The bus ride the next day made me appreciate how long and narrow Chile really is. The only main highway in Chile runs from one end of the country to the other, with the ocean on one side and the mountainous Andes on the other. On this road, buses are the principal means of transportation; made for serious long-distance travel, they offer air conditioning, comfortable seats, toilets, and snack and drink dispensers. I was quite impressed.

Four hours later we arrived in Coquimbo, and I realized we had gone a mere fraction of the 2,500-mile length of the country. The northern part of Chile is extremely arid, with some areas never having seen rain. The countryside around Coquimbo, though not quite that dry, reminded me of some of the drier regions of the California desert.

The project architects, biologists, and I spent one busy week working on the aquarium and came up with a design for a nice, small aquarium. It had one large community tank and a number of small- and medium-sized tanks for the display of Chilean fishes and invertebrates.

The biology students at the university were friendly, enthusiastic, and knowledgeable, and one evening I was invited to one of their homes for dinner. Even in the supposed privacy of their home, there was only whispered talk of political happenings. It was a country of fear and oppression. During my stay in Coquimbo the mayor of the city was gunned down in the lavatory of City Hall. The official newspaper reported it as a suicide, but everyone knew that he'd been an outspoken critic of the government and his death was one more political assassination. The heavy hand of the dictatorship was evident everywhere. I kept my mouth shut.

Upon entering the living room of the university student, I was bowled over to see Earl Ebert, a California Fish and Game biologist I'd dived with many years ago while I was working at Marineland. He was now head of California Fish and Game's Granite Canyon Lab south of Monterey, and his specialty was the culture of abalone, the giant, edible mollusk of California. He was in Chile to help set up a culture operation for the red abalone (*Haliotis rufescens*).

CHILEAN BARNACLES

Dr. Cea graciously made sure I got my diving wish when, lending me a wet suit, tank, and regulator, he took me scuba diving. It was fascinating. Although the water temperature was similar to southern California's, ranging between 55°F and 68°F, the marine life was strikingly different. Underwater I noticed that the surfaces of all the rocks were almost completely covered with small barnacles. They were so dense, there was little room for anything else.

I remembered reading somewhere that the sailing route from Ecuador heading due west was notorious for boats and ships becoming fouled with barnacles. This route crosses the Humboldt Current flowing up from Chile in the south. There, covering every rock, I saw the reason behind the complaint. Those millions of Chilean barnacles release their billions of planktonic larvae into the northward-flowing current—swarms of barnacle larvae just waiting for a nice boat hull to come along on which to settle and grow.

The rocks around Coquimbo, at least in the areas where I dove, were almost bare of algae. Although Chile is known for its rich growth of seaweeds, these are found farther south, where the water is colder. Instead of plant life, almost every crevice between the rocks was crawling with one- or two-inch-long red shrimp. When I stuck my hand in between the rocks, the little shrimp crawled on it as if they were cleaning me. I was struck by the rather ominous thought that if you should die in this water, you would be picked apart by hordes of these hungry little crustaceans.

Northern Chile has a commercial diving fishery for a mollusk the natives call *loco* (*Concholepas*), also known, inaccurately, as the Chilean abalone. The loco looks more like a giant limpet, but unlike limpets or true abalone it is carnivorous and feeds on the abundant barnacles. I felt sorry for the loco divers I saw. Although the water temperature isn't very cold, their wet suits were tattered and full of holes. Not only that, but they were using hookah diving, where the air to the regulator and mouthpiece is supplied by a compressor on a small boat rather than by a self-contained air tank; because their air supply is unlimited, hookah divers can—and do—stay down for hours. In addition to hypothermia, then, these divers ran the risk of the bends. There were also occasional white shark attacks on the loco divers. It looked like a hard life.

TINY EASTER ISLAND

As a thank-you for my help with the university's aquarium design, Dr. Cea gave me a return flight to San Francisco via Easter Island, Tahiti, and Hawaii, with the option of staying as long as I liked in each place. He'd spent a few years on Easter Island as the sole medical doctor, he said, and he loved the place. He told me of a dream he had to build a public aquarium there. I listened politely, not wishing to disillusion him by pointing out that this island, the most remote inhabited spot

The giant carved heads, or *moai,* of Easter Island. (Photo by author)

on earth, with a total population of two thousand and very few visitors a year, was the last place one should build a public aquarium.

Leaving Santiago on Lan-Chile Airlines, I arrived on tiny Easter Island. The island has three names, depending on your nationality. The Polynesian inhabitants know it as Rapa Nui; the Chileans, whose flag flies over it, call it Isla de Pascua; and the English refer to it as Easter Island. Famous for the massive carved stone figures, or *moai,* it has a fascinating history.

The remote island was colonized at least once and possibly twice in the past by Polynesian seafarers who may have strayed far off course and by chance made a fortunate landfall before they died of thirst and starvation in the vast South Pacific Ocean. The closest land is Pitcairn Island, 1,600 miles away. Although there was not a single tree on Easter Island when discovered by the Dutch captain Roggeveen in 1722, archeological evidence indicates that the island was once covered with dense forests.

The history of the island from the arrival of the original Polynesian inhabitants (around A.D. 400) until 1722 reads like an ominous warning. When first colonized by humans, life on the island must have been good and food plentiful. The people thrived and the population grew to an estimated ten thousand. Things did not remain peaceful, however. Competition between tribes grew fierce, with the creation of the giant carved heads becoming a form of rivalry. The already decimated forests were cleared for logs to roll the massive stone heads from where they were carved to their chosen resting places several miles away.

With the forest gone, the climate changed and food became scarce. War and fighting were now the way of life, and the population declined drastically. The once-productive island was changed forever. It struck me that the history of this tiny island is a microcosm of what's happening in our world today. I only hope that we can recognize and control our destiny before we, too, follow the path taken by the colonizers of Rapa Nui. It must have been far from pleasant.

The message seems clear. If ten thousand people can do that to one island, what will six billion—or more—do to our island in space, Earth? The similarities are striking. Just as on Easter Island, our renewable resources are being used up faster than they're being replenished, food is becoming scarce in many parts of the world, forests are being cut down, climates are changing, and there's competition among the "tribes." The big question is: Will we be able to stop before our "last tree" is destroyed?

Scuba air was not available here, but I'd brought my fins and mask and was able to snorkel at several spots around the island. I was struck by how sparse the marine life was, both in the variety of species and in total numbers. The island, which drops off steeply into the surrounding deep sea and has no coral reefs, stands in clean oceanic water that is poor in nutrients and plankton; as a result, productivity is very poor. It must be very difficult for the islanders to live off the sea.

SPECTACULAR TAHITI

I left Rapa Nui on the twice-weekly Lan-Chile flight to Tahiti, having arranged to stay at the home of an American scuba-diving fish collector. Using what little French I remembered from my school years back in England, I took one of the open-air taxis from the airport to his house on the beach south of the capital of Papeete.

I was shocked to find my host crippled with the bends. Although he could walk laboriously with crutches, he spent much of his time in a wheelchair. Discussing his accident, he told me that he'd been diving deep trying to catch a new species of fish. He spotted what he recognized as a new species of angelfish at a depth of over two hundred feet. Driven by his desire to capture this new fish, he pushed himself beyond the limits of safety and went even deeper. His downfall, however, was that he had not allowed adequate time since his last deep dives for his body to rid itself of dissolved nitrogen.

Decompression sickness, or the bends, is caused when nitrogen gas that has dissolved in your tissues—at depth and so under high pressure—forms bubbles when you rise too quickly to the lower pressure at the surface. It's just like uncapping—letting the pressure off—a bottle of carbonated soda water. These bubbles can become lodged in critical parts of your body and cause a variety of symptoms, from joint pain to paralysis and even death. If a person rises to the surface slowly enough to allow the dissolved nitrogen to gradually leave the body through the lungs, clinical symptoms don't occur. Making repeated deep dives too close together, however, results in nitrogen accumulating in the tissues, with a much greater danger that bubbles will form on ascent. That's exactly what happened to my friend.

He still wanted to dive, though, and as long as someone was along to help him he was able to do so, but he stayed in water shallower than sixty feet. We went on several dives together; he said the aches and pains he experienced on land disappeared as soon as he was underwater.

The island of Tahiti is truly spectacular, with its precipitous green mountains, lush valleys, and palm-fringed beaches. I found the native Tahitians friendly and cheerful, but the French I met were just the opposite—many were rude, unfriendly, and mercenary. The prices for the simplest items in the French-run shops were incredibly high.

Even more spectacular than the land was the underwater world. The coral reefs and their hosts of inhabitants were healthy and, to my inexperienced eye, seemed unaffected by the nearby activities of man. Unlike Easter Island, Tahiti was abundant with marine life.

The productivity of the photosynthetic coral reefs makes for an incredibly rich and complex ecosystem. Tuna, jacks, and gray reef sharks patrolled the outer reef, and whitetip reef sharks were sleeping the day away under overhanging table corals, waiting for nightfall, when they could come out and look for food.

RAIN IN RAROTONGA

My four days in Tahiti passed too quickly, and soon it was time to catch my flight to Hawaii and home to California. Checking in with the airline, I was told that Pan Am had discontinued its service from Tahiti to Honolulu. They therefore rerouted me from Tahiti to Rarotonga in the Cook Islands, where four days later a Qantas flight would take me to Hawaii—at no extra charge. I loved it! Luckily, I was in no particular hurry to get back to Steinhart. John McCosker probably wouldn't even notice I was gone, I thought optimistically as I looked forward to another adventure in a new country.

I arrived in Rarotonga in a pouring rainstorm and learned that a big weather front had just moved in and rain was expected for the next three days. That was disappointing news. Pouring rain is not the best condition in which to see a country; it would, I feared, put a literal damper on my short time on this beautiful island.

After finding a small hotel just outside of town, I started asking around about diving possibilities. There was no dive shop on the island, but I was told that the "boys from the crash crew" at the airport were all divers. I looked them up and, sure enough, they were more than happy to take me diving the next day. It seems they went diving every chance they had, even in the rain. Because of the infrequency of flights to Rarotonga, they had plenty of free time.

The next day we met at the airport firehouse and drove to the little harbor where they kept their boat. We headed out and dropped anchor just a short distance from the harbor. I donned the tank and regulator they loaned me and we went down. Was I ever in for a shock! I'd expected to see something like the reefs at Tahiti or Canton Island, but instead found only massive coral heads of several different types that were completely dead. There were dead brain corals as big as small cars. Growing in isolated spots on these heads were little fist-sized corals, apparently started by larvae that had settled out of the plankton. I saw some fish swimming around these stark reefs, but not in the abundance typical of a healthy, living reef.

I asked the crash crew divers what had caused this massive die-off, but they didn't know. They'd come to Rarotonga fairly recently from New Zealand to work at the new jet airport, and whatever killed the coral had happened before they arrived. To this day I have no idea what caused the die-off. Theories abound, including pollution from agricultural chemical runoff, typhoon damage, or upwelling of deep, cold water, but these are all just guesses. The coral will eventually grow back, but it may take centuries before it reaches the massive size of some of those dead brain corals.

The steady rain eased up a bit on my last day, and I took advantage of the lull to rent a car and drive around the island. I found a historical marker which proclaimed that at that very spot many hundreds of years ago seven oceangoing canoes of the "Great Fleet," led by master navigator Kupe, had embarked on an epic migration southward to Aotearoa, "land of the long white cloud." It was an incredibly risky gamble. Who knew if the new home to which they headed even existed? Without the use of a compass, they navigated by deciphering the prevailing waves and clouds and reading the paths of the stars and of migratory seabirds.

Those pioneering people became known as the New Zealand Maoris, the first settlers of that island nation. Some in Polynesia dispute this story and say the Maoris came from elsewhere in the South Pacific, but the Cook Islanders firmly believe they came from Rarotonga. What were the circumstances leading to that heroic voyage? Was their island becoming overpopulated, its resources depleted and the land no longer able to support the numbers of people? It's interesting to speculate on their motivations, and one must surely marvel at their skill and endurance to survive such a journey.

My short time on this beautiful island over, I caught the Qantas flight to Hawaii and then home. I would love to revisit Rarotonga before it becomes changed too much by the spreading influence of Western civilization.

15

MONTEREY BAY AQUARIUM

In 1978, reports began circulating that plans for an aquarium in Monterey were being hatched. A number of Academy staff asked me about it, but I knew nothing more than what the rumor mill was churning.

Then I got a phone call from Julie Packard in Monterey. She told me that, indeed, serious plans were afoot for a new aquarium that would focus entirely on Monterey Bay marine life. She asked if I would consider coming to Monterey to review and discuss their plans. I agreed to take a look.

A few days later I drove to Monterey. That first meeting was held in a low wooden building at Stanford University's Hopkins Marine Station, next to the old Hovden sardine cannery at the end of Cannery Row. I was quite familiar with the area, having stored my boat in the ramshackle Monterey Canning Company building next to Ed Ricketts's lab some years before. I'd also made countless dives in the 1960s and 1970s for Steinhart Aquarium, Sea World, and my own pleasure in the waters around Monterey.

At the meeting were the originators of the project: Nancy Burnett, Julie's sister; her husband, Dr. Robin Burnett; Dr. Steve Webster; and Chuck Baxter. Julie and Linda Rhodes, the project manager, were also there. Chuck was teaching marine biology at Hopkins, and Nancy, Robin, and Steve were marine biology graduates of Stanford University. Julie's master's degree in biology, earned with a thesis on marine algae, was from the University of California at Santa Cruz across the Bay. I was impressed with their enthusiasm and their extensive knowledge of the invertebrate and algal life of Monterey.

They seemed a bit weak on fishes, however, and they freely admitted that they had little experience actually keeping animals in aquariums. They told me they needed help from someone in the business, so to speak. The plan, an ambitious one to be sure, was to exhibit all of the environments and microhabitats of Monterey Bay, from a living kelp forest all the way down to the meiofauna, those almost microscopic organisms living among the grains of sand. It would be the first major aquarium in the United States to focus exclusively and in depth on local species displayed in natural communities. Virtually all other major aquariums displayed species and environments from many different parts of the world.

THE BAY'S RICH PROMISE

I could sense the group's enthusiasm for the rich marine life of Monterey Bay and their desire to share it with others. That was just the way I'd felt when I started diving in the 1950s, and I still felt the same. Robin, Nancy, and Steve had all taken invertebrate zoology from Don Abbott at Hopkins. It's interesting how a truly inspiring teacher can have such an impact upon your life. I was fortunate to have had such a teacher in Boyd Walker for ichthyology when I was at UCLA, but wished I'd also had an equally inspiring mentor for invertebrate zoology and marine ecology.

The four instigators—Robin, Nancy, Chuck, and Steve—told me they'd dreamed up the idea of an aquarium during an evening of lively conversation inspired by liberal consumption of margaritas. The original idea was of a small combination aquarium-coffeehouse on Cannery Row, but the more they talked, the more they wanted to include—and, of course, the more ambitious the project became.

Even a modest public aquarium is not an inexpensive undertaking, and the subject of how to fund it came up. They decided to approach Nancy and Julie's parents, David and Lucile Packard, to see if they might be talked into funding it. David Packard was chairman and co-founder of the Hewlett Packard Corporation. He could certainly afford to finance it if he could just be convinced that it was practical.

However, before jumping into what might be a harebrained scheme in a field he knew little about, Packard commissioned the Stanford Research Institute to do a feasibility study. The conclusion of the study was favorable, indicating that a small aquarium on Cannery Row could be financially successful and might expect an attendance of around 350,000 a year. With this good news the enthusiasm of the aquarium's planners grew, and so did the scope of their ideas.

The main exhibit was to be a large roundabout similar in concept to the new Fish Roundabout at Steinhart. As with the Roundabout, the visitors would be in the center and the aquarium would surround them. Within the large tank would be re-creations of the major subtidal environments of Monterey: the kelp forest, deep granite reefs, shale beds, and the sandy seafloor. The animals and plants found in those environments—giant kelp (*Macrocystis pyrifera*), sea otters (*Enhydra lutris*), rockfish, sea perch, lingcod (*Ophiodon elongatus*), flatfish, sardines, and a host of invertebrates—would all live in this large, unpartitioned exhibit.

The assumption was that because certain animals are found in well-defined environments in Monterey Bay, they would stay in the appropriate environment in the roundabout. Unfortunately, I knew that wouldn't work. With predators and prey in the same exhibit, the predators would seek out or ambush potential prey and in a short time the only ones left would be the ones with the biggest mouths. For a few weeks, of course, we would see the concept of survival of the fittest—and the hungriest—in action.

The three major problem species were the sea otters, the lingcod, and the salmon. Sea otters, being the intelligent and curious mammals they are, would quickly eat everything they found edible and then would destroy much of the rest of the exhibit through their curiosity and playfulness. I expected that they would pretty much trash the exhibit within a month. Meanwhile, the lingcod, well-camouflaged ambush predators, would methodically eat every passing fish that would fit into their large mouths while the speedy salmon picked off the schooling anchovies one by one. I could see why the founders needed a little help from someone who understood the behavior of animals in aquariums.

My predictions of the carnage that would occur in a single all-encompassing exhibit were heeded, and the one large roundabout idea was tossed out in favor of three exhibits. This decision would allow the predators and troublemakers to be kept apart from their prey and from more fragile habitat components, like the living kelp.

On my next trip to Monterey I looked at the new three-part configuration, which consisted of a kelp forest, a sea otter exhibit, and one large tank featuring four different environments of Monterey Bay: the deep granite reefs, the sandy bottom, the shale beds, and the wharf pilings. I recommended that if they wanted to have good species representation of the smaller fishes and invertebrates, these needed to be displayed separately from their potential predators. The logical division was to put the smaller fishes in the Kelp Forest and the larger predatory fish in the Monterey Bay Habitats exhibit.

The interaction between predators and prey may be a part of the natural world, but it's not something we want in an aquarium. Apart from the loss of valuable specimens, we like to think that our charges will live a long life without the fear of ending up as lunch. As a result, compromises often have to be made in the design of live exhibits.

An example of a successful compromise was the wharf piling habitat, a key component of the Monterey Bay Habitats exhibit. In the real world of the ocean, a wide variety of small fishes would be able to survive in such an environment, where predatory lingcod and large-mouthed rockfish like the bocaccio (*Sebastes paucispinis*) rarely naturally go. In an aquarium, however, these small fishes would not last long, even in a very large tank providing nooks and crannies for cover, with such predators lurking nearby.

The solution was to display full-sized wharf pilings in the Monterey Bay tank, to suggest the overall look of the habitat, and then, in a separate, smaller exhibit, place invertebrate-covered sections of pilings and smaller fish species, like the live-bearing surfperches (Embiotocidae) that are so abundant beneath the Monterey wharf. Not only would this make life easier—and longer—for the little guys, but it would also give visitors a close-up view that they would otherwise not have had.

It was at this second meeting that I met the architect, Chuck Davis of the San Francisco firm Escherick, Homsey, Dodge, and Davis. Working with the biologists, Chuck had come up with preliminary designs for the three new exhibits. The Kelp Forest, for example, was a tall, twenty-eight-foot-deep tank that would be perfect for the giant kelp—assuming, of course, that we could grow kelp at all, something that had never been done. I noticed, however, that the orientation of this exhibit was such that the afternoon sun would shine directly on the main windows, which would both reduce visibility and significantly increase diatom and algae growth. Luckily, the design of the building was still in the concept stage and changes were relatively easy to make. After calculating the sun angles at different times of day and year, Chuck rotated the entire tank so the sunlight would strike the back wall and not the windows. I was impressed with his awareness and flexibility. It was a pleasure to work with an architect who considered the exhibits and the animals of highest importance and understood that the role of the building was to support and highlight them in the best way possible.

A DREAM JOB

Over the next few months I made more trips to Monterey to meet with the group and review the changes and developments in the exhibit program and the building design. Although some of the proposed exhibit ideas were impractical or possibly boring, I began to be intrigued with the goals of the new aquarium and the potential for creating really interesting displays.

One factor that profoundly affects the richness and diversity of live exhibits is the threat of a fish disease—specifically, a parasitic ciliated protozoan called *Cryptocaryon irritans,* the bane of any aquarium displaying tropical and subtropical fishes. The most effective treatment has been the daily addition of copper sulfate in precisely measured concentrations, which can kill one stage of the protozoan but not the stage when it's embedded in the skin of the fish. This resistant stage makes it extremely difficult to permanently rid a water system of the parasite. Moreover, copper sulfate is quite toxic to invertebrates and plants. This means that exhibits of fishes that also include invertebrates are often impractical because if the disease crops up, all the invertebrates must be removed in order to treat the fish.

Like many aquarists, I had my share of frustrations at Steinhart and Sea World battling this stubborn disease. I did learn one thing in my skirmishes over the years with this ciliate, however, and that is that it is unable to reproduce below 63°F. If the water system is colder than that, there will be no *Cryptocaryon* infecting the fishes.

Knowing this, I saw the tremendous possibilities for mixed-species exhibits at Monterey. The water temperature in Monterey rarely gets as high as 63°F, a fact that would allow us to display fishes plus a host of delicate invertebrates and marine algae all together in the same exhibit. Slow-growing invertebrates and seaweeds could be left undisturbed for years to thrive and develop.

The more I thought about it, the more I realized that this could be a great opportunity to create aquarium exhibits that truly represent the diverse fish, invertebrate, and algal life of the ocean. This was the vision of the four founders, and I also saw it as an opportunity to follow up on the idea I'd had twenty years before at Marineland of putting aquarium backgrounds out in the ocean to become naturally encrusted with marine growth. Left out long enough, they would come to resemble the real thing.

My meetings at Monterey led to an offer of the position of curator for the new aquarium. I was still a little nervous about being able to pull off some of the proposed exhibit ideas, but the potential was certainly exciting. I accepted and was asked to come to Monterey for a meeting and interview with David Packard. After some preliminary conversation about the aquarium, Packard asked me to join him in the other room. It was a job interview, but for the life of me I don't remember what was said. It must have gone all right because, with a smile, he said I had the job, shook my hand, and told me the salary I would receive.

This was the spring of 1980, and our two daughters had grown up and were now pretty much out on their own. The move to Monterey wasn't such an ordeal for Betty and me, except for the usual hassle of selling a house and finding another place to live. After several weekend trips from Marin County to Monterey to house-hunt we found the perfect place in Pacific Grove, known as the "Butterfly Town" because it is a winter home of monarch butterflies. The property had no lawn (I have a thing about growing lawns in the semidesert of California), a large garden for vegetables, and a place to park my ski/fishing boat and camper, and the house had a nice big kitchen.

I remembered that a few years earlier Betty and I had driven to Monterey and Pacific Grove with my sister, Marcia, and her husband, Sir Guy Lawrence, who were visiting from England. We'd mentioned to them that someday we wanted to retire here. Well, here we were in 1980, living on the Monterey Peninsula. Far from being in retirement, though, I was helping attend the birth of a most ambitious new aquarium.

16

CREATING THE EXHIBITS

After settling into our new home, I began full-time work at the Monterey Bay Aquarium. The habitats and the stories we wanted to tell about Monterey Bay were gradually taking shape, and design of the tanks was under way.

I was charged with deciding on the size, shape, and placement of most of the live exhibit tanks. Being familiar with the requirements of the animals, I was able to specify the living space and environment that would meet their needs. I also played a role in determining the look and feel of the exhibits from the public's point of view.

An experience I'd had at Steinhart had a major influence on me and in turn on the design of the Monterey exhibits. Shortly after the 1964 renovation of the Steinhart Aquarium I was standing in the public corridor in front of the tanks watching the reactions of visitors as they came to each exhibit. I wanted to see which exhibits and which animals most interested them.

In setting up the new displays at Steinhart, I'd made sure that the rockwork, sand or gravel, and plants, as well as the animals, in each tank were as different from those in adjacent tanks as possible. My aim was to make each exhibit unique. Yet watching the visitors, I was struck by the relatively short time most of them spent at each tank. Some people simply started at one end of the corridor and kept walking until they reached the other end. They would glance into each tank, but they never stopped walking.

I noticed, from where I stood at the beginning of that long hallway, that the outsides of the tanks all looked the same. Every window was the same height, width, and distance from the ground. Nothing on the exterior of all those exhibits differentiated one from another. It dawned on me that for some people the identical size and shape of the outside of the tanks subconsciously meant that things were much the same on the inside too. There was little motivating them to stop and really look at what was on the other side of the glass.

The first-glance impression is a key element in determining a visitor's overall day-at-the-aquarium experience. With that observation in mind, I knew I had to make the outside of every exhibit tank at the Monterey Bay Aquarium appear distinctly different. The aquarium windows must vary in height, width, size, and sometimes shape from any others within sight. This gives the visitor the subconscious message that if the outside of each exhibit looks different, then the inside must be different and therefore worth stopping to see. I believe that approach has worked, and, except on impossibly crowded days, most visitors do stop and look at what it is we're trying to show them.

LIVING KELP FOREST

Many exhibit projects needed to be tested to see if they would work. The most experimental exhibit of all, however, the proposed living kelp forest, was something of a gamble from start to finish.

A spectacular and unique underwater wonder of the world is the giant kelp forest habitat of the California coast. Not only was the growing tip of the giant kelp chosen as the aquarium's logo, but this beautiful plant would be our centerpiece exhibit as well. Our vision was to create for the aquarium visitor the exhilarating feeling experienced by a scuba diver "flying" weightlessly and freely through a forest of gently swaying golden kelp plants.

But there was a problem. Until now, only small-scale experimental work had been done with giant kelp; no one had ever grown it to any significant size. We talked to people who knew kelp the best, having worked with it in either the field or the laboratory—experts such as Wheeler North of the California Institute of Technology and Mike Neushal of the University of California at Santa Barbara. They gladly gave us all the information they had on the nutrient requirements, water flow velocities, and light intensities that make for happy kelp. They also wished us a lot of luck.

Relying on their information and our own educated guesses, we proceeded with the design of the tank. There was really no point in testing anything on a small scale; that had already been done. The acid test would be when everything was finished and full-sized kelp plants were brought in and tried out. We had fallback plans, of course. If the kelp didn't do well, we could continually replace it, or, as an absolute last resort, we could use the realistic-looking plastic kelp that Carl Gage of Biomodels had developed. None of us relished this last thought. Our goal was to bring the natural world to the visitor, and we wanted every part of it to be real.

Some other exhibit ideas, however, could be tested and would give me the information I needed to design the best exhibit for a particular animal or subject matter. One such experiment was to find out whether sand dollars (*Dendraster excentricus*) would behave in an aquarium the same way they do in the ocean.

Just outside the breakers along certain sandy beaches live vast colonies of sand dollars. Most of these spiny, flat animals stand on edge, looking as if someone had come along and stuck them in the sand like rows of silver dollars. It almost looks artificial, but that's how they feed, trapping particles of organic debris that wash by them in the back-and-forth motion of the wave surge. Their upright position in the sand is entirely functional.

I tested their behavior in a circular fiberglass tank, using water currents of varying velocities and directions. Eventually I found a combination that resulted in many of the sand dollars standing up on edge just as they do in the real world.

Exhibit designer Jim Peterson and I then tackled the challenge of creating an interesting display with an animal that has no color, doesn't move, and lives on a featureless sandy bottom. Potentially, this exhibit was a real snoozer. However, putting into practice my notion of variety in tank shape, we drew up a design: an eye-catching quarter sphere of clear acrylic that has not only worked for the sand dollars, but has also lured people into stopping, looking, and learning.

THE PROGRAM TAKES SHAPE

The overall exhibit program was coming together. Some ideas were dropped, such as trying to show animals that live beneath rocks. That one became a logistical nightmare: the animals simply didn't like light and either disappeared or became very unhappy when they were exposed to lights. While they might have been quite happy if we'd let them stay under a rock in the dark, that wouldn't have made for much of a display.

Once the exhibit program was more or less in place, finalizing the exact sizes, shapes, and placement of the smaller gallery tanks became a high priority. This needed to be done quickly so the architects could keep on schedule designing the layout of walls and drains, the electrical system, and climate control for the building, all of which had to be coordinated.

I began by sketching the tank sizes and shapes that I felt best met the needs of the animals as well as reflected my ideas on how to display them. I used the empirical method of cutting out pieces of paper to scale and moving them around on the plan until everything fit. I changed the sizes and shapes of the tanks slightly here and there where necessary, but in the end this method worked rather well. I made sure there was room for holding tanks, utility sinks, aquarists' tools—and people. I then drew the final tank shapes and sizes on the floor plan, which was given to the architects. These days this would all be done with a computer and a CAD program. It may have been low-tech, but hey, it worked.

Typical of all pioneering projects, not everything went smoothly. There were many false starts, and some ideas were simply abandoned. One disaster involved an attempt to make artificial rocks. The Rock and Waterscape Company in Los Angeles had been contracted to make artificial FRC (fiberglass-reinforced cement) rocks that we could place on the ocean bottom to be colonized by marine life. The rocks were

AN UNEXPECTED DISTRACTION

Early in 1981, with opening day still three-plus years off, the entire Monterey Bay Aquarium staff consisted of nine employees, and we were all busy focusing on our particular tasks.

Julie Packard and Steve Webster spent a great deal of time spreading the word in the community about the up-and-coming aquarium. Showing a well-used set of slides, they gave countless talks to just about any civic group around the Monterey Peninsula that wanted to hear about their project. They wore many hats during this time, chief among them being marketing and public relations.

Together with Robin Burnett, Chuck Baxter, and Steve, I was compiling and finalizing the species lists, developing appropriate tanks for each exhibit, and designing the seawater systems. I was also starting to think about the monumental challenge of collecting and caring for all these creatures.

Our focus was suddenly disrupted with the arrival of an unexpected visitor, an orphaned sea otter pup that someone had found on the beach. Even though we didn't exist yet as a real aquarium, with real seawater or real tanks, apparently the community considered us a real aquarium and therefore we must be the place to take a baby otter.

There are few creatures in this world as appealing as a sea otter pup, and we were all captivated by it. Unfortunately, only one of us, Dr. Tom Williams—our veterinarian, who was helping design our sea otter exhibit—had ever seen an otter pup close up, and even his experience with young otters was limited.

Experienced or not, our hearts were stolen by this furry little bundle, and we would do anything we could for it. Until now, however, virtually all rescued sea otter pups had died, with the notable exception of one raised by Dean and Bertha Tyler at the Morro Bay Aquarium.

Although the odds were against us, we all hoped that because this was *our* otter, it would survive. He was a male, about a month or two old, and he was already able to take small pieces of solid

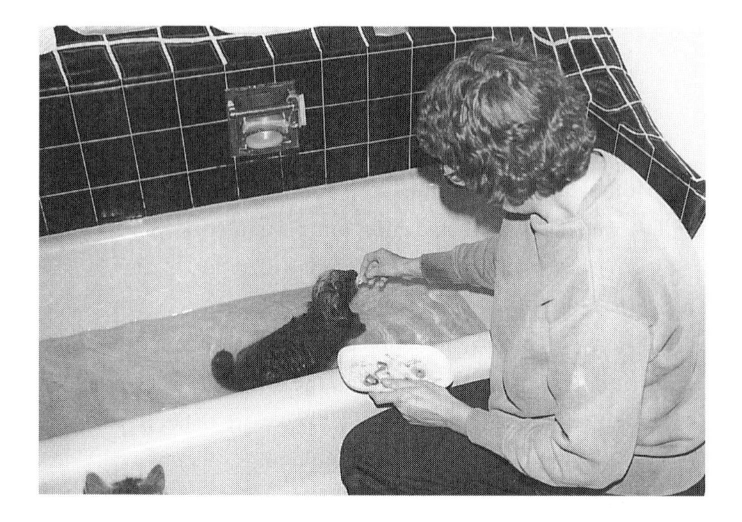

Betty Powell feeding the aquarium's first sea otter pup in the bathtub at home. (Photo by author)

food like clams, squid, and shrimp. (Sea otters have expensive tastes in food!) We bought a plastic toddler wading pool, which he floated in like a furry cork. Typically, of course, as soon as he was put in the little pool, he would poop and the water had to be changed. Tom had warned us about letting his fur get soiled or too wet, so he would be taken out of the pool and dried, first with towels and then with a hair dryer. Tom also warned us about the tendency of sea otters to overheat because of their dense fur.

Unless he was asleep he spent much of the time letting us know with a piercing scream that something or other was not to his liking. As a result, not a lot of work got done while he was at the office. Because sea otter pups require around-the-clock care, at the end of the day he went home with one of us for the night. There wasn't a lot of sleeping done when it was your turn. I remember Julie Packard and Linda Rhodes coming in bleary-eyed several mornings after their turn with the pup.

The nights the baby otter spent with the Powells will never be forgotten. He spent hours bathing, pooping, being dried with towels and a hair dryer, and crying. Betty and I, meanwhile, dutifully carried out our many otter chores, spelling each other as needed. Much of this activity was accompanied by his charming scream.

At the time we had a wonderful, gentle Sheltie at home, and she would grow quite agitated and concerned whenever the otter screamed. Maternal by nature, she tried to respond to his cries, knowing he needed help but having no idea just what she was supposed to do for this strange, demanding creature.

Those long nights impressed upon me the very real survival function of an otter pup's loud scream. Earsplitting though it is in the confines of a small room, in the otter's watery world it's imperative that it be heard. A mother spends much of her time foraging underwater for food—both to sustain herself while she's nursing and, later, to share with her hungry pup—and often comes up some distance from where she left her pup floating on the surface. At water level of a frequently choppy sea, she can seldom see where her tiny pup is floating. Instead she relies on its piercing cry to guide her back. Along the shores of the central California coast you can often hear the cries of pups as they await the return of their diving mother.

Sad to say, our first sea otter pup died, certainly not from lack of trying but because of our own ignorance. Three years later Tom Williams, Pat Quinn, Julie Hymer, and a dedicated staff of otter volunteers successfully raised four orphaned pups to adulthood for our sea otter exhibit, learning much along the way. Today, the rescue and care of stranded sea otters, while never a sure thing, is marked by considerably more successes than failures. We do learn from our mistakes.

---

quite large, but they were hollow and so not very heavy for their size. Marine biologist Dr. Gene Haderlie of the Naval Postgraduate School in Monterey was on our board of directors, and he arranged for us to use the research vessel the *Acania* to take the rocks out for placement on the Pinnacles in Carmel Bay.

The Pinnacles are impressive granite peaks that rise out of deep water to about eighty feet from the surface. Because of the exceptionally rich life covering the rocks, we thought this location would be ideal for our first rocks. We figured in three years they would be covered with growth and we'd go back before the aquarium opened to retrieve them. We had a boatload of people on board for this momentous event as we headed around to Carmel Bay. Some were divers, while some were just along for the ride and to watch the launching of our first fake rocks.

When we arrived at the Pinnacles, we hoisted the first rock off the deck with the ship's crane and lowered it into the water. There was a bit of a swell running that day, and as soon as the rock was underwater the roll of the ship combined with the drag of the water broke the rock in two. That wasn't a good beginning for our ambitious project.

Indeed, it quickly became obvious that the rocks were far too weak to survive what we had planned for them. During severe winter storms the Pinnacles are scoured by heavy ground swells, the effect of which can be felt as deep as a hundred feet. Our rocks would surely be swept away and broken into many pieces. Even if that didn't happen, they were so light underwater that when giant kelp spores settled on them and grew, the plants' buoyancy would eventually lift the rocks off the bottom, and kelp, rocks, and all would drift away to who knows where. When a piece of our rocks was found many months later, it already had a number of small kelp plants growing on it. We learned several lessons from that failed venture: the rocks had to be very strong; they needed to be placed in an area protected from storm waves; and they needed to be solidly bolted down if they were to stay where we put them for three-plus years.

THE FINE ART OF NATURE FAKING

David Packard took a personal interest in the way the project was developing, and it wasn't just because he was paying the bills. As a self-made man who, together with Bill Hewlett, started the Hewlett-Packard electronics company in his garage, he had a do-it-yourself philosophy. He also believed that if a project was going to be done, it should be done right. However, Packard didn't get upset if something didn't work out at first. There are lessons to be learned from first mistakes that point you in the direction of getting it right.

We held many discussions about how to get all the ambitious plans for the new aquarium from the idea stage to reality. In line with his do-it-yourself attitude, Packard said we would set up our own fabrication facility to create components of the exhibits. To head up this venture he hired Derek Baylis, an ingenious Australian-bred engineer who had considerable first-hand experience fabricating a wide range of products using a variety of materials. Among other accomplishments, he'd built his own sailboat and sailed it from Australia to Tahiti.

David Packard and Derek Baylis rented a large warehouse in nearby Sand City for the shop where this creative work would be done. To assist him with this challenging project, Derek then hired Andy Anderson and a number of skilled craftsmen who had been building sailboats across the Bay in Santa Cruz.

Derek's responsibility was to create one-of-a-kind exhibit elements. His "to do" list included the fiberglass aquarium tanks and the artificial rockwork habitats inside the tanks—for although the rockwork for the three largest exhibits, the Kelp Forest, Monterey Bay Habitats, and sea otter exhibit, was to be made by the Larson Company of Tucson, Arizona, we would be fashioning the natural rock environments for the smaller, focus exhibits ourselves. Derek's crew would also manufacture an array of marine mammal models, ranging from six-foot dolphins to a forty-two-foot gray whale mother and her calf; build wave-generating machines; and fabricate the wall panels in which the aquarium tanks were to be set.

There was considerable discussion about what material would be best for these panels. They had to be fireproof, seawater proof, sound absorbent, and also movable so we could make exhibit changes in the future. After much research, Derek settled on the relatively new fiberglass-reinforced cement for both the wall panels and the artificial rockwork. Unlike conventional cement, FRC is reinforced with glass fibers and can be made relatively thin and light without sacrificing strength. It has been used extensively in Europe for building wall facades. The process of making FRC involves a special machine that chops and blows alkali-resistant glass fibers at the same time that it forcefully sprays a thin cement slurry. Much experimentation was done by the Sand City shop crew, and ultimately they produced beautifully made wall panels.

Chris Anderson, who had experience with rocks from prospecting in California's Gold Country, took on the task of developing techniques for the manufacture of artificial rocks. We needed two kinds of artificial rocks. For ones that needed to look real because they would be close to the viewing public, latex molds were made of actual granite rocks that had interesting surface textures. These molds were then used to form the surface texture of the FRC rock, which ended up looking and feeling just like the real thing. For rocks that we expected would be quickly overgrown by algae, detailed surface texture was not important, so no latex molds were required; shapes and rough texture were created by carving the cement before it became hard.

After some testing, the rock-making techniques were perfected and the Sand City shop went into production. We began by making artificial rocks that fit exactly inside the fiberglass exhibit tanks. The first exhibit we worked on would show the visitor the differences in the animals and plants that grow on vertical versus horizontal rock faces in the kelp forest just offshore. To achieve this, it was important that the rocks be in the ocean as long as possible before the aquarium opened so they would have plenty of time for plants and animals to settle and grow on them. Acting on the lessons we'd learned from the failure of our first flimsy rocks, I designed a heavy steel structure that we secured to the seafloor with several hundred pounds of concrete. To this structure we would firmly bolt the FRC rock panels.

Marine life that grows on hard surfaces in the sea faces intense competition for living space as constant "space wars" are waged between organisms. In these battles, each organism uses potent chemical "weapons" to keep interlopers at bay. The result of this jostling for space is that the distribution of animals becomes uneven as certain animals become dominant and successfully fight off invaders. One of the goals of this exhibit was to show this dominance and the resulting patchiness of distribution.

Just putting our artificial rock panels out in the ocean would eventually result in their becoming well encrusted with some form of marine life or other, but there was no telling if it would be an animal we wanted to exhibit. We came up with a rather sneaky way of helping nature along to make sure that each panel would be dominated by the animal we wanted. Just prior to moving the rock panels out to the steel rack, marine biologist Linda Martin and I experimented with gluing a few choice animals, like sponges and sea anemones, onto the FRC surfaces using a nontoxic marine epoxy that cures underwater. The animals survived, and once they were put out in the sea they grew and multiplied. Because we had given them a head start in the space wars, they eventually took over most of the rock surface.

We made periodic dive inspections on the rocks to monitor the progress of the animal's growth and to evict predatory sea stars. Ultimately, the patchy distribution and dominance of certain animals we wanted to show our visitors was achieved, thanks to our little partnership with nature.

FUNKY WHARF PILINGS

One big, early project was to make artificial wharf pilings, which we would then attach to real pilings underneath the wharf in Monterey Harbor. These pilings were destined for the small exhibit of wharf-dwelling fishes I mentioned in chapter 15. Layers of fiberglass cloth impregnated with black-pigmented polyester resin were laid over a smooth vinyl form to create six-foot-tall cylinders, which were then reinforced on the inside. We took them out in the boat and dived to bolt them to the animal-covered pilings of the wharf.

At the time, no opening date had been set for the aquarium, so I didn't know how long we would have for animals to colonize the bare fiberglass. As it turned out, the pilings were in the sea for four years. By then they were so covered with growth that they looked just like the real pilings of the Monterey wharf.

The Monterey Bay Habitats exhibit, however, was to be fitted with real pilings, and we were hoping to get them from the Monterey Harbor when the old pilings were replaced, which happens periodically. That way they would be suitably funky looking—just what we wanted.

Our hope was that the Monterey Harbor Department would replace pilings shortly after we had our exhibit tank filled with seawater (projected for the summer of 1984) and we'd be able to move them directly into place. Unfortunately, that wasn't to be. We heard from the harbormaster that they'd be pulling some pilings out in early 1983 but then not again for several years. This was a bit of a problem: we had pilings when we didn't need them and there'd be no pilings when we did need them.

The harbormaster was more than happy to give us the ones they pulled in 1983 instead of cutting them up and hauling them off to the landfill, but what would we do with a dozen animal-covered, fourteen-foot wharf pilings in the meantime? After a little brainstorming, we came up with a solution: we'd secure them to the existing pilings underneath the wharf until we were ready to move them into our exhibit. Fortunately, the very cooperative harbormaster thought that was a fine idea.

Granite Construction, which was contracted to do the piling replacement, used a crane to pull the pilings out of the muddy bottom of the Bay and hoist them onto the wharf to be cut up and hauled away. This time, however, we were waiting there, and as soon as we saw a piling covered with an attractive assortment of live creatures, we asked for it. The workers then cut it to the exact fourteen-and-a-half-foot length we needed for our exhibit and tossed it back into the water.

Some of the pilings were waterlogged and sank straight to the bottom. Others were still slightly buoyant and floated. Using our small Boston Whaler, we towed the pilings one at a time under the wharf and, with a block and tackle, winched the buoyant pilings twenty feet down to the bottom. The sunken pilings we raised upright using diver's lift bags filled with air. Each piling was then tightly secured to an existing piling with stainless steel bands called Band-It. It was both a time- and scuba air–consuming job, but by taking turns over several days we managed to secure ten nice animal-encrusted pilings. Over the coming months we made periodic dives to check on our stash of pilings and to re-band any that loosened in the ever-present surge beneath the wharf.

A year and a half later the Monterey Bay Habitats exhibit was finally full of seawater and ready to become the new home for our pilings. The trick now was to get them out from under the wharf and into the exhibit without injuring the animals growing on them or, for that matter, us.

Getting ready to bring the first wharf pilings in to the Monterey Bay Aquarium. (Photo © 2000 Monterey Bay Aquarium Foundation. All rights reserved.)

We dove down and selected the eight best-looking pilings from the ten we'd originally saved. After removing the stainless steel bands we attached floats to each end of the first one and slowly towed it over to the boat-launching ramp. We backed the boat trailer down into the water and floated the piling on top of it, being careful not to crush any animals. The piling was then covered with sacks wet with seawater to keep the animals alive for the short drive down Cannery Row to the aquarium. There, it was hoisted up, lowered into the Monterey Bay tank, and towed over to its final position.

The laborious task of moving one piling at a time took several days and the combined work of the two collectors, eight aquarists, and myself. It was a tricky job, but the final result looked fantastic—almost like it does when you dive under the wharf. The big difference was that the water in our exhibit was a lot clearer than the often murky plankton-rich water in the harbor.

The wharf piling end of the Monterey Bay Habitats was located next to an inlet pipe that would supply unfiltered seawater each night to bathe the filter-feeding animals with plankton. This would also be periodically supplemented by volunteer and staff divers feeding them a "krill shake" from a squeeze bottle, a rich, nutritious slurry made from krill whipped up in a blender.

THE SHALE LOOK-DOWN

Among the more interesting underwater environments of Monterey Bay are the shale beds. Located just east of Monterey Harbor, these soft sedimentary rocks jut a few feet out of the sandy bottom of the Bay and are home to rock-boring clams, or piddocks (*Zirphaea* spp.). The shells of these clams can reach a length of five inches, and as the animals grow, they enlarge their holes by grinding their rough shells against the soft shale. With their shells and bodies totally concealed in the rock, all that is visible is the flowerlike tip of their siphon lying flush with the rock's surface.

When these clams die, from old age or by being eaten by a predator, a cavity is left in the rock that's quickly taken over by another animal. The shale rocks therefore are home to many more animals than are found on the hard surfaces of granite reefs. As the shale is eroded by waves and bored by animals, flat slabs break off and fall to the bottom below. Still filled and covered with animals, these slabs of shale can be collected and brought into the aquarium.

Many years ago I dived off Los Angeles and collected shale to keep in my home saltwater aquarium. The abundant life in those rocks fascinated Betty and me. For this reason, one of my favorite displays at the Monterey Bay Aquarium is the shale look-down exhibit. It is an open, shallow, eight-foot-long tank, the bottom of which we covered with slabs of "living" shale. On the surface of the water are floating magnifying glasses, which the visitor can move around to get an up-close-and-personal look at the creatures living on and in the rocks. My only regret is that we weren't able to make the exhibit a little larger. Once the aquarium opened, we found that visitors tend to lose themselves in these shale menageries while others impatiently wait their turn.

17

AQUARISTS AT WORK

DESIGN OF THE AQUARIUM was progressing well and it was now time to hire key staff members for the husbandry department. Skilled people are not easy to find, so putting a competent staff together would mean luring them away from other aquariums. Fortunately, the world of public aquariums is small, and I was acquainted with most of the experienced aquarists on the West Coast and knew their situations.

I had met Chuck Farwell, assistant curator of the Scripps Aquarium in La Jolla, ten years before when I was at Sea World. I knew he wasn't too happy with the existing situation at Scripps, so I called to ask him if he'd be interested in joining me in Monterey. Chuck jumped at the chance; he also said that he could highly recommend Mark Ferguson, a truly dedicated aquarist who had started out as a volunteer at Scripps and was now working as a senior aquarist. Both Chuck and Mark were hired, and they moved up to Monterey within short order.

I needed one more experienced aquarist and thought immediately of Michael Weekley from the Seattle Aquarium. Prior to going to Seattle, Mike had been an aquarist at the Waikiki Aquarium under director Bruce Carlson. He had specialized in the collection and husbandry of a deep-sea cephalopod, the chambered nautilus (*Nautilus pompilius*).

Filling the job of collector was easy. Bob Kiwala and I had worked together on several adventurous and successful collecting trips while he was at the Scripps Aquarium. I definitely wanted him to join us in Monterey.

Bob, however, was in Rarotonga in the South Pacific. Several years before, while participating in a Scripps Institution of Oceanography scientific cruise that stopped in the Cook Islands, he'd fallen in love with the South Seas paradise. When the cruise ended, Bob packed his hammock, a few books, his cigars, and belongings and flew back to Rarotonga. He rented a hut at the far end of the island for ten dollars a month. It was idyllic and, knowing Bob, he must have fit in well with the Cook Islanders. He met a young Cook Island woman and they had a baby daughter.

I sent Bob a letter telling him about the exciting new aquarium in Monterey and asked if he would become the chief collector. He said he would—lacking a head for money matters, he may have discovered he couldn't live on papayas, fresh fish, and love alone—but then he kept putting off the date when he would actually show up to begin the job. Considering the lifestyle he was leaving behind in Rarotonga, I really couldn't blame him for stalling.

Eventually, Bob made his way back to the States, moved to Monterey, and started work. I'd hoped he would arrive in time to choose the type of collecting boat and equipment he'd be using, but we had to get started collecting the thousands of animals we needed by opening day. I just couldn't wait for him any longer, so I went ahead and purchased a trailerable, twenty-five-foot fiberglass Farallon boat. In honor of our benefactors we named the boat *Lucile* after David Packard's wife.

I'd already prepared a list of all the species we would need, with approximate numbers, and I'd prioritized the animals according to hardiness, natural longevity, and difficulty collecting them. We would begin with those fishes that would do well in holding tanks until the exhibit tanks were ready for them. At that point, those first fishes would be moved to the exhibits, making room in the now vacated holding tanks for different species. It would be an ongoing fish-juggling act.

Critical to the success of this plan was a holding area that would be fully functional well before the seawater system in the main building began operation. The space beneath the employee parking deck was deemed the best location for our holding tanks. An important criterion was that they be shaded from sunlight, both to minimize algae and diatom growth in the tanks and to avoid stress to the fish.

The intake pipes that brought water into the pump house from beyond the kelp bed had just been completed. A temporary submersible seawater pump was installed in the sump of the pump house and a pipeline run from the pump around the perimeter of the building construction site back to the holding tanks. It was an exciting day when, at last, power to the pump was turned on and we all gathered in an atmosphere of celebration to watch the first tank fill. Somewhat preoccupied with all the preparations, I was unaware that some of my coworkers had hatched a devious plot. In one swift, well-coordinated motion my colleagues grabbed my arms and legs from behind while my treacherous wife, Betty, whipped the wallet out of my back pocket, and I was tossed, clothes and all, into the tank. Wet and salty though I was (and not for long, since Betty had thoughtfully brought along a set of dry clothes), it really was a fitting way to celebrate the beginning of this new phase for the aquarium, a phase where we would be working with real animals instead of blueprints, pictures, and lengthy to-do lists.

The temporary water system worked well. We didn't have twenty-four-hour staff at the aquarium yet, so to alert us to problems with the water supply a pressure switch was installed in the pipe, and it automatically dialed the phone number of a security company if the pump stopped. The security company would then call us. Luckily, we got only a few middle-of-the-night calls to go to the aquarium to check out the pump and the fish.

BRINGING IN THE ANIMALS

By the time the aquarium opened we would need thousands of fish, representing hundreds of species. For some of them I imposed on old friends, such as Steinhart Aquarium director John McCosker and curator Tom Tucker, who helped us immensely by giving us striped bass (*Morone saxatilis*) and white sturgeon (*Acipenser transmontanus*) from San Francisco Bay. There's an unwritten understanding in the public aquarium world that we help one another whenever we can. No records were kept, but the Steinhart staff knew that we would repay them in the future when they needed a favor.

Although I had experience collecting most of the animals on the species list, there was one fish I had neither collected nor kept in an aquarium. The king, or Chinook, salmon (*Oncorhynchus tshawytscha*) is a key commercial and sport fish in Monterey Bay, and it was important that we have some nice large ones on exhibit.

I had seen very few large salmon in other aquariums and I worried about how the fish would react to capture and quarantine in our twenty-foot-diameter holding tank. I suspected they might be flighty and possibly jumpers; it turned out that, at first at least, they were, and we installed jump screens to prevent their escape.

Reading everything we could on salmon fishing and talking to salmon fishermen convinced us that the best collecting method for our purposes was slow-speed trolling. The wire line on each heavy rod and reel would troll four leaders with barbless hooks and plastic squids, known locally as "hoochies," as lures.

Not quite sure what we were doing, we ventured out into Monterey Bay. We fumbled around with the gear for a while until we got the hang of it, and we were delighted when we actually caught some salmon. Practice makes perfect, and we got better at it over time. I knew that salmon have fine scales and that a net would knock them off, breaking the skin's protective mucus layer and leading to infection. To protect their delicate skin, we never touched the fish with a net. Instead, we would lift the fish with the strong leader straight from the sea into the holding tank on the boat.

Upon returning to the harbor, the boat was winched up onto its trailer, towed to the aquarium, and backed down right next to the holding tanks. The salmon were then moved one at a time in smooth, strong plastic bags from the boat to quarantine.

Powerful but delicate fish require careful handling. For several weeks the salmon would startle anytime they saw someone approach. To get them to feed, we'd sneak up to the side of the tank and toss anchovies or herring in without them seeing us. I always wondered if their pronounced startle behavior from seeing a movement above the water had evolved from millions of years of encounters with grizzly bears fishing for salmon as they returned to the streams to spawn. In time our fish became comfortable with people and eventually would swim over to greet us when we brought food.

DEAD AND DYING SQUID

Although most of the exhibits were based on the habitats of the Bay and the animals and plants found in them, there was one small section that focused on a group of especially interesting animals. The gallery of cephalopod mollusks would feature the giant Pacific octopus, two species of smaller octopus, the chambered nautilus from the South Pacific, and the local species of squid (*Loligo opalescens*). At least, that was the plan. Our local squid, however, had never been kept for any length of time in aquariums, and I wasn't at all sure we could do it either.

Squid come into the Bay by the millions at the end of their second year to reproduce. Aggregating in huge schools, they mate, lay their finger-shaped white egg capsules—which come to look like a vast shaggy carpet on the sandy mud bottom just beyond the kelp beds off Cannery Row—and die. It's an orgy of sex, egg laying, and death.

Although easy enough to collect when they come in to spawn, the squid, male and female alike, are close to the end of their life. If we could collect the juveniles, they would have a longer life, but it's still a mystery where they go after they hatch from the millions of eggs that are laid every year in Monterey Bay.

Faced with this dilemma of the squid's natural life cycle, we collected some adult squid just before the opening of the aquarium. They made for a fascinating but somewhat shocking (and short-lived) exhibit as the squid mated, laid their egg capsules, and went through their death throes. A major drawback was that the bottom of the tank was littered with dead and dying squid. This wasn't easy for visitors to understand, and I'm sure a lot of them left thinking that a catastrophe had occurred.

Another drawback was that after three weeks we had no more live squid. So we wouldn't be stuck with an empty tank, we temporarily switched the exhibit to a school of anchovies, and a year later we introduced another cephalopod, the cuttlefish (*Sepia officinalis*), from Europe. At least it fit with the cephalopod theme of the gallery, even though this species, like the adjacent chambered nautilus, doesn't occur in Monterey Bay.

The chambered nautilus still needed to be collected. Senior aquarist Mike Weekley asked if he could take part of his vacation time plus some work time to assist Dr. Peter Ward of the University of California at Davis with an ongoing research project investigating the biology of the chambered nautilus off the coast of New Caledonia. While he was there he would collect some nice large adult specimens for our exhibit. It sounded like a good opportunity for him, for us, and for science.

But then tragedy struck. Two weeks after his departure I received a phone call informing me that Mike had died in a diving accident. I was stunned. The woman on the phone could provide no details, but we found out later that Mike and Peter Ward had been diving quite deep and Mike had fallen unconscious well below the hundred-foot depth. He never regained consciousness.

The first suspicion was that the air in their scuba tanks had been bad, but that was later ruled out by French officials in New Caledonia. No autopsy was done, and even if it had been I'm not sure it would have pinpointed the cause of death. We do know that in order to work with the chambered nautilus they were making many deep dives, and the cumulative physiological effects of repetitive deep dives can be dangerous. What made it all so upsetting was that Mike was young, strong, and at the beginning of a promising career.

One of the hardest things I've ever had to do was go to Mike's house and tell his wife, Jamie, what had happened. His body and personal belongings were shipped back to Hawaii for burial. Mike's death occurred in 1984, and I still find myself thinking about him and how things might have been if he were still with us. It was a tremendous loss to all who knew him.

At the Waikiki Aquarium, where Mike had his first job as an aquarist, and at the Monterey Bay Aquarium, the exhibits of nautilus each bear a plaque: "In honor and memory of Michael Weekley." I know Mike would be proud that people can now see living nautilus, a species whose exhibition he helped pioneer.

By February 1984 construction of the building had progressed to the point where non-construction workers like us were finally permitted access. Now we could start setting tanks in place and connecting them to the water systems. In preparation for this day we had added staff to the husbandry department. Most were brought on in the position of all-around-aquarist, although some were specialists.

Roger Phillips, who had a degree in phycology (the study of marine algae), was responsible for our centerpiece exhibit, the Kelp Forest. When preparing his aquarium business cards, the printer assumed Roger had misspelled his title, so he "fixed" it: "Staff Phycologist" became "Staff Psychologist." Just in case Roger ever decides to change his profession and give us all psychotherapy, he has his business cards ready.

Bruce Upton and Randy Hamilton were hired for their experience with the culture of oysters and abalone. Pat Quinn came from the SPCA Wild Animal Rehabilitation Center; she had primary responsibility for the sea otters and the shorebirds. Julie Hymer's past experience with mammals at a zoo prepared her well for her work with sea otters. Freya Sommer, coming to us from Hopkins Marine Station right next door, was highly recommended for her knowledge of local marine invertebrates. And finally, Neil Allen joined us to be the assistant to our collector, Bob Kiwala.

Outsiders often wonder what exactly aquarists do. Feed the fish? Well, yes, but there's a whole lot more. The number one requirement of a good aquarist is to have a fascination with nature and, in particular, a love of animals and a feeling for what they need. This means, simply, that an aquarist must have an indefinable ability to "read" an animal and to tell just by looking how it's doing.

Beyond that, it is the responsibility of an aquarist to set up and maintain interesting, healthy exhibits that provide accurate glimpses into the lives and environments of aquatic creatures. This simple-sounding goal involves a wide variety of essential tasks.

First, the creatures in an exhibit, whether they're mammals, birds, fishes, invertebrates, or plants, need to be obtained—either by being collected, purchased, traded, or cultured—and transported with minimum trauma back to the aquarium. This process can vary from relatively easy to incredibly difficult. It is a common misconception that collectors, or aquarists who do their own collecting, are people who get paid to go fishing or diving. Well, they do, but fishing and diving is the easy part. The real challenge is handling the animals gently, keeping them alive and healthy, and getting them back to the aquarium in good condition. It can be quite nerve-wracking knowing a life is in your hands.

Most new animals need to be treated for parasites or disease before going on exhibit. An aquarist needs to be able to calculate aquarium volumes and the correct dosage of therapeutic drugs. Accuracy is vital; an error of a decimal point can be fatal. Animals must also be feeding well before they are taken out of quarantine. Figuring out and getting the right foods for animals that have specialized needs are an important and often challenging part of the aquarist's work as well.

Much of the work of an aquarist is cleaning—cleaning glass, cleaning walls, cleaning gravel. This is important not only for the health of the animals but also for aesthetic reasons. We want the aquarium, with its necessary walls and windows, to disappear as much as possible. Even something as minor as a tiny patch of algae growing on the glass or on a tank wall will likely catch the eye, reminding the visitor that this is actually a man-made tank with animals in it. The less of the actual container that is readily apparent, the more easily visitors will mentally transport themselves into the world beneath the water and focus their attention on the plants and animals within.

Some of the work consists of hard physical labor. Fishes and invertebrates live in water, and water is heavy. Just moving animals from one place to another often means moving heavy buckets or tanks of water. Likewise, washing and hauling wet sand and gravel is strenuous. We've all been told that the correct way to lift heavy objects is with the legs and not with the back. Well, that may work for some jobs, but it's impossible to lift "correctly" while leaning over the edge of a tank and lifting out a large rock.

An aquarist's work can be repetitive, and to some it may seem menial, but the knowledge that thousands of people come to see one's exhibits is excellent motivation to keep them in top shape. Much of the reward of this work comes from watching and listening to visitors as they react to what they're seeing and experiencing in the watery world of the aquarium.

THE HOME STRETCH

The day came when the complex plumbing and electrical systems that were to sustain life for all of our future creatures were complete and the big intake and recirculation pumps were turned on. We filled the Kelp Forest in June 1984, less than five months before the aquarium's October opening date. Seawater started pouring into the exhibit at a

---

VOLUNTEER DIVERS

In addition to a staff of sixteen full-time aquarists, the Monterey Bay Aquarium has about 500 volunteer guides, or docents, whose role is to help visitors understand what they're seeing in the exhibits and to add more meaning to their visit.

Although we'd anticipated using some volunteer divers as well, we were quite surprised by the large number of people who offered to help out. We now have a crew of 120 divers—and there's a long waiting list—who perform much of the underwater maintenance of the large exhibits. This involves the careful cleaning of the soft, easily scratched acrylic windows, siphoning of debris from the bottom of the tanks, and twice-daily underwater feeding shows in the Kelp Forest exhibit.

Having dived many times in Monterey Bay, I knew well how cold it is. I was concerned that long and repeated exposure to that water would have a chilling effect on the enthusiasm as well as the bodies of our volunteers.

We couldn't do much about warming up Monterey Bay, of course, but we could do something about warming up the divers. The best way to recover after getting chilled to the bone is

The spectacular experiment: the Kelp Forest exhibit at the Monterey Bay Aquarium. (Photo © 2000 Monterey Bay Aquarium Foundation. All rights reserved.)

through immersion in hot water. I wanted to get a hot tub for the volunteers, but I worried that nondivers would consider my request as a decadent California luxury. I nevertheless went ahead and got bids on the cost, then submitted my request for the euphemistically named "thermal recovery unit," with a brief explanation of why it was needed. I waited for some reaction.

A while later my wife, Betty, happened to be riding down the administration building's elevator with David Packard. He turned to her and said with a grin, "Tell David he's got his hot tub!" I was delighted to hear that Packard had seen the value to the aquarium of taking care of our invaluable volunteers. The hot tub is used frequently by the volunteer divers and has been the site of a number of post-dive pizza parties.

rate of 2,000 gallons per minute—fast enough to fill a bathtub in three seconds. In three hours the 310,000-gallon tank was full. Excitedly we all went up to the top of the tank to look down into its twenty-eight feet of crystal-clear water.

For weeks I had planned to celebrate this day by pushing Julie Packard into the freshly filled exhibit. When the time actually came, though, I noticed that she was wearing a really nice dress and I chickened out. She was standing right at the edge; one little nudge would have done the job. As a phycologist, she would have been the most fitting person to "baptize" in the centerpiece exhibit of our brand-new aquarium. I bet she would have felt honored, too.

There are precedents for such apparently disrespectful acts toward bosses in the aquarium industry. We threw Sea World's president George Millay into the pool at a party to celebrate the end of our first successful summer, and Baltimore's mayor jumped into a tank at the National Aquarium when that aquarium failed to open on the promised day. If only Julie had come to work in blue jeans that day. . . .

The few months leading up to the opening of the aquarium were incredibly busy. Although Bob and Neil were the official collectors, everyone helped out. It's important that all the aquarists participate in the collection of the animals, rocks, and plants they will be caring for in their exhibits. Not only do they learn new skills, but they also appreciate their charges much more once they realize how difficult they can be to collect. Just getting out and seeing the environment where these animals live also gives the aquarist a picture of how an exhibit should look and what it is we're trying to show our visitors.

18

COLLECTING THE FISH

VISITORS TO AQUARIUMS have certain expectations about what they want, or at least hope, to see. At the top of most people's list are whales and dolphins, octopuses, eels, and of course sharks. It was decided at the very beginning of the Monterey Bay Aquarium project that we wouldn't have cetaceans. Not only was public sentiment growing in the United States and Europe against keeping highly intelligent, social animals like dolphins and whales in captivity, but Sea World, with its large marine mammal exhibits, was already doing an excellent job of exhibiting orcas and bottlenose dolphins.

We would, however, have the Pacific giant octopus and the scary-looking but actually docile wolf eel, as well as plenty of sharks collected from Monterey Bay and Elkhorn Slough, an estuary at Moss Landing, midway between Monterey and Santa Cruz. Leopard sharks and gray smoothhounds (*Mustelus californicus*) are perfectly fine sharks, but most of them are small—two to four feet long on average. And small is not what most visitors have in mind when they think of sharks. They want to see big sharks, sharks that fit the Hollywood prototype; some even want "Jaws" itself. Of course, that portrayal is mostly fantasy. Still, I wanted at least to show them sharks of a respectable size.

The only large shark that is at all common in Monterey Bay is the blue shark, but I knew from past experience that they would never survive in the Monterey Bay Habitats exhibit. Although the tank is quite large, this open ocean shark wouldn't be able to tolerate the obstructions of rocky reefs and wharf pilings.

Great white sharks, while uncommon in the Bay, periodically make their presence known by a rare but spectacular attack on a surfer or seal or a gray whale carcass. However, even very small white sharks have failed to thrive in aquariums. In any event, the only great whites that frequent Monterey Bay are large adults of a thousand pounds or more, and they're much too big to even consider trying to bring in alive. Just because an animal is large and dangerous does not mean it is tough. In fact, quite the opposite is true. The sheer size and weight of such an animal makes moving it a monumental job, which in itself becomes life-threatening to the animal. Moving such a massive animal is not particularly dangerous to the aquarists, but it most certainly is to the shark.

SEVENGILLS IN HUMBOLDT BAY

I knew I had to come up with a shark that was both large and indigenous to Monterey Bay. The logical choice was the sevengill (*Notorhynchus cepedianus*), an animal that lacks the typical sleek, streamlined shark shape and has been stuck with the most unglamorous—and inappropriate—family name of cowshark.

Sevengills certainly don't have much in common with docile, herbivorous cows. They do get large, however, and they do have a big mouth. Their teeth are well concealed within their mouth, and in their normal swimming mode they look as if they have no teeth at all, like an old man who forgot to put in his false teeth. In reality, those teeth are quite formidable and with little effort can take large chunks out of prey. In the shallow, muddy bays that sevengills inhabit, bat rays are one of their favorite foods, and they've also been known to eat harbor seals. They are top predators; they just don't advertise it with their appearance.

Sevengills, though found in Monterey Bay on occasion, are not common there. I had, however, collected a number of them, along with soupfin sharks, in San Francisco Bay for Steinhart's Fish Roundabout. The sevengills are found in San Francisco Bay on a fairly regular basis, but the soupfins were occasional and highly unpredictable visitors. Unfortunately, while I was at Steinhart a sport and commercial fishery for sevengills had opened up, and in a relatively short period of time there was a noticeable decline in both their numbers and their sizes.

Clearly, if we wanted a good chance of collecting large sevengills, we needed to look elsewhere. Dave Ebert (son of Earl Ebert, whom I had dived with during my Marineland days) had just completed his master's thesis on the biology of sevengill sharks in California and South Africa. He put me in touch with Ken Bates, a commercial fisherman from Eureka, California, who had helped Ebert gather data for his study.

I called Ken and then, in June 1984, flew up to Eureka to check out the logistics of collecting sevengills there. Ken said that during the summer months large sevengills six to eight feet in length show up regularly in Humboldt Bay. He'd caught them then and had seen them chasing bat rays over the shallow mud flats. The situation looked very good for collecting. Ken had a fine herring gillnet boat he had built himself that we could use. Additionally, a convenient dock was available for transferring the sharks from his boat to our trucks. The marine lab at Humboldt State University helped by agreeing to let us fill our transport tanks with their nice clean, filtered water rather than using the murky water of Humboldt Bay and then having to filter it ourselves.

The following month, Bob Kiwala, Neil Allen, aquarist Gilbert Van Dykhuizen, and I drove up to Eureka in two trucks. The morning after we arrived we headed out with Ken in his boat to put out the setlines in Humboldt Bay. He used a method similar to one I'd used with Gerry Klay in the Florida Keys to collect lemon and nurse sharks: each line consisted of a single weight for an anchor, a fifteen-foot leader and baited hook, and a line going up to a float.

We'd brought boxes of frozen salmon heads for bait. The heads were split in two and half a head was put on each of fifteen lines. Returning in an hour or so, Ken pulled each line in to see what we'd caught. There were a couple of five-foot leopard sharks, which we released; one line, though, had just what we were looking for: a nice six-foot sevengill. Maneuvering the shark into the stretcher alongside the boat, we lifted it on board and lowered it into the long, plastic-lined "shark coffin" transport box. The oxygen pump was then plugged in to bring the dissolved oxygen above saturation level. Just as it had for Sea World's blue sharks, this high concentration of oxygen would compensate for the lack of water flowing over the gills of the sevengill.

All this was done with utmost care. Sevengills are quite flexible and can move quickly when they want to, and they need to be handled with a lot of respect. To let down your guard when they're quietly lying in the stretcher could be a tragic mistake. Their razor-sharp teeth are, after all, designed to cut through flesh.

That first day of fishing we caught two nice large sevengills, which we transferred to the tank on one of the two trucks. We then called the aquarium to let them know the good news and to alert them to be ready for us when we returned. That done, Gil and I took off for the ten-hour drive back to Monterey, stopping every hour or so to check the sharks' condition and to make sure oxygen was flowing from the pump.

Fortunately, our timing was just right and we got through the San Francisco Bay area after rush hour traffic had subsided. Creeping through bumper-to-bumper traffic with a tank of delicate animals can definitely test your frustration tolerance. Arriving at the aquarium shortly before midnight, we were happy to see Chuck Farwell and three aquarists waiting to help move the two sharks into the Monterey Bay Habitats exhibit. We backed the truck under the electric hoist, and the three-thousand-pound transport tank with its two precious occupants was lifted up and inside the building.

On the deck adjacent to the exhibit, the lid to the tank was unbolted and removed and the tank lifted again, swung out over the water, and lowered in. At that point the lifting bridle was unhooked from one side of the transport tank, while the other side was lifted gently to allow the two sharks to swim out. No doubt confused and stiff from being confined in the tank for over ten hours, the sharks took a little while to become oriented. Until they were able to maneuver flawlessly, the husbandry staff gently fended them away any time they looked as if they might bump into the walls or the windows. Fortunately, the two aquarists who were standing by with wet suits and scuba gear weren't needed to help the sharks swim.

Two days later Bob and Neil called to say they had two more sevengills and would be coming in that night. Those sharks also did well. All in all, the trip was a great success. Three months before the aquarium was scheduled to open, we had four healthy, large sevengills cruising around the Monterey Bay Habitats.

The next step was to get them to take food. None of the current husbandry staff had ever worked with sevengills or, for that matter, any large sharks. What's more, and not surprisingly, having these large, potentially dangerous animals in the exhibit raised a certain amount of apprehension on the part of the staff and volunteer divers who would be cleaning the windows, siphoning the bottom, and feeding some of the invertebrates. At Steinhart I'd dived a number of times with the sevengills in the much narrower Fish Roundabout, which had only nine feet between the window and the wall of the tank. I'd found it quite safe as long as I left the shark alone and didn't crowd it.

Partly to feed the new sharks, but also to show others that it really was safe to dive with them, I decided to try a method I'd used before to get food to new sharks, especially when they're outcompeted by speedy, aggressive eaters like the yellowtail at Steinhart. Carrying a short wooden pole with a dead mackerel impaled on the end of it, I positioned myself at the narrow part of the hourglass-shaped Monterey Bay tank and waited until the first sevengill came by. I then thrust the food in the path of the approaching shark so that it would pick up its scent. My ploy worked: the shark opened its mouth, grabbed the mackerel, and gulped it down whole.

Many sharks, like the mako and the blue shark, depend on their keen eyesight to provide information about the world around them. Others, however, such as the tropical bull shark, the sixgill (*Hexanchus griseus*), and the sevengill, live in dark, murky water where even good eyesight isn't of much use. These sharks depend instead on an acute sense of smell, which tunes them in to both living and nonliving food; on the ability of their lateral line system to detect vibrations coming

The four sevengill sharks did quite well, and it wasn't long before the staff and volunteer divers felt comfortable sharing the tank with them. Although the design of the exhibit tank successfully showed visitors the inhabitants of the four different habitats of Monterey Bay, it wasn't ideal for sevengills, who normally spend little time around rocky reefs.

The sevengills had a problem at feeding time when food such as a whole mackerel fell to the bottom. Using its acute and highly directional sense of smell, a sevengill would zero in on it and go down and try to pick it up. Because much of the bottom was rocky, the sharks suffered abrasions on their snouts. They would also occasionally bump the walls while turning around.

Unfortunately, due to repeated bumping, the white-colored abrasions of the nose skin on some of the sharks didn't heal. We eventually decided to release them back to the ocean and replace them with new ones collected in Humboldt Bay.

On one such collecting trip for new sevengills in July 1990, we put out the baited setlines just as we'd done on previous trips, hoping luck would bless us on our first day with a nice six-footer to take back to Monterey. When we checked the lines a couple of hours later, one of them had something on it that definitely didn't want to come to the boat.

It was quite a struggle to finally get the animal alongside. The catch turned out to be the largest sevengill that any of us, including sevengill-experienced Ken Bates, had ever seen. It looked huge and was far too heavy for the onboard crew to lift into the shark box. Ken secured the shark alongside, and at idle speed he headed over to where some oyster harvesters were working in shallow water. By enlisting their cooperation and muscle, we managed to lift the shark on board and safely into the box.

The shark was a female, about ten feet long with an estimated weight of around three hundred pounds. She was close to a world-record size for sevengills and was certainly the largest ever in an aquarium. Ken ran the boat back to the dock, and the shark was carried in the shark stretcher to the transport tank on one of the trucks. With lots of oxygen pumping into the tank, aquarists Gilbert Van Dykhuizen and Scott Nygren took off for Monterey.

Ken headed back out to check the remaining setlines and found there was a nice seven-foot male on one of the other lines. This one was much easier to bring on board and was transferred to the transport tank on the second truck. It had been a great day. We had two sharks, and one of them was the largest ever seen in an aquarium.

Ten hours later both sharks were swimming around the Monterey Bay Habitats exhibit. The big female looked huge. Not only was she long, but she was so big in girth that we were convinced she must be pregnant. She began eating quite soon and did well, avoiding the sides of the tank and the rockwork and wharf pilings. Although the aquarium had a policy against giving names to our animals, she was unofficially christened "Big Emma," named after the ample proprietress of a local diner.

As time went by she became less rotund and pregnant-looking, and after a year we concluded she definitely wasn't going to give birth. It's quite possible that she'd originally been pregnant but had absorbed the eggs or developing embryos.

Over the next four years Emma did well, but there were times when she displayed her predatory nature, which was a bit of a worry. On several occasions she bit and even consumed some of her tank mates. Most disturbing were the deep slashes she made on our prized white sturgeons. Fortunately, our veterinarian, Tom Williams, was able to suture the wounds, and the sturgeons recovered completely.

By 1994 Big Emma had developed a white abrasion on her snout and was also showing disconcerting, possibly aggressive behavior toward divers in the tank. She would swim over a diver and stop almost motionless overhead. This, of course, was most unnerving.

We decided, for her benefit and for the good of the fishes and divers who shared her tank, that we'd release her into Monterey

"Big Emma," the ten-foot sevengill shark, impresses the crowd. (Photo © 2000 Monterey Bay Aquarium Foundation. All rights reserved.)

Bay. She was herded over to the side of the exhibit and maneuvered into the same transport box she had arrived in four years earlier. The transport tank was then lifted out, lowered into the collecting boat, *Lucile,* and taken out into the bay.

Emma was tagged with an external identifying tag and released. On October, 16, 1996, in Humboldt Bay, two years and four months after her release in Monterey, a sportfisherman captured Emma. She had returned to the very same bay she came from six years before. We were sad to hear that she'd been killed, but it was heartening to know that an animal like that can survive quite well when released back to the wild.

Over the years several aquariums have released a number of fishes and sharks. Unlike on land, where visibility is good and radio-tagged animals can be followed, we've rarely known how well these released fishes have survived. We've always assumed they did fine because we thought they'd readily revert to their instincts for survival, but we've rarely had proof.

In the case of Emma, with her external tag and subsequent recapture, we have positive proof that not only did she survive, but she thrived. In addition, she undertook a truly remarkable migration—about four hundred miles over territory she'd probably never traveled by sea. After all, she came down to Monterey by road in a truck, so she certainly didn't have any recognizable underwater landmarks on her way back to Humboldt.

Over the years biologists have come to understand quite a bit about the anatomy and physiology of animals. However, we know very little about the subtleties of their behavior and we understand even less about the remarkable migrations that some animals make. Their ability to travel great distances to specific locations without the use of a compass, clocks, and maps is something that puzzles us technology-minded humans. We're still baffled by the amazing migration of third-generation monarch butterflies many hundreds of miles from Canada back to the very trees in Pacific Grove, California, where their grandparents spent the winter.

There are many other almost unbelievable examples, such as the return of the Pacific salmon to the stream of their origin after spending three years wandering and feeding in the ocean. Or the return of the adult eel (*Anguilla*) from Europe across the Atlantic Ocean to its spawning grounds in the Sargasso Sea. We can now add Big Emma the sevengill shark and her kin to this list of remarkable migrators.

---

to them through the water; and on their ampullae of Lorenzini, specialized organs for detecting the tiny electromagnetic fields produced by the contracting muscles of all living creatures. The potential prey doesn't have to be actively swimming to give off these electrical fields; the beating of the heart is enough. (Except for one other very strange animal, the duck-billed platypus, this electrical sense is unique to sharks.)

We found out later that sevengills can be enticed to eat by using the same fish-on-a-stick technique, but with the feeder standing on the deck above the water. That was certainly much easier and less time-consuming.

It's a common misconception that sharks are ravenous eating machines waiting to devour anything they come across. In fact, nothing could be further from the truth. A long-term study done by Gilbert Van Dykhuizen and volunteer Henry Mollet on the food intake of our sevengills has shown that they have an extremely efficient metabolism and need surprisingly little food—the equivalent of just 0.2 percent of their body weight daily—to function and grow. In this regard, sharks make an interesting and striking contrast to sea otters, which require 20 to 25 percent of their body weight daily just to survive.

SANTA CRUZ ISLAND

The main message we want our visitors to carry home with them is that the marine life they see in our exhibits is actually found right here in Monterey Bay. Now, that doesn't necessarily mean that we collected all the animals in Monterey Bay. It does mean, however, that they're known to live here, or at least that they travel through the area. In fact, many animals may be uncommon or only occasional visitors to the Bay, being much more abundant elsewhere.

From the practical standpoint of collecting, it always makes sense to go where the species can be found in large numbers. With this in mind, we planned a trip to Santa Cruz Island, twenty miles offshore from Santa Barbara and two hundred miles south of Monterey, to collect certain fishes for the Kelp Forest exhibits.

Ernie Brooks, owner of the renowned Brooks Institute of Photography in Santa Barbara, offered to help. In addition to teaching commercial photography, the Brooks Institute offers a thorough course in all phases of underwater photography, and Ernie has a fifty-seven-foot boat, *Just Love,* which he uses for the practical phase of these classes. The boat comes complete with cabins, bunks, and galley for overnight stays at the nearby Channel Islands—perfect for use as a floating base camp. The plan was to trailer our *Lucile* down to Santa Barbara, launch it, and run it over to the island to rendezvous with Ernie and the *Just Love.* The target species for this trip were fishes that are not common in Monterey, such as blacksmith fish (*Chromis punctipinnis*), halfmoons

(*Medialuna californiensis*), horn sharks, swell sharks, garibaldi (*Hypsipops rubicunda*), sheephead (*Semicossyphus pulcher*), ocean whitefish (*Caulolatilus princeps*), and kelp bass (*Paralabrax clathratus*).

The morning after arriving in Santa Barbara we launched the *Lucile* and headed for Santa Cruz Island, where we would rendezvous with the *Just Love*. Halfway across the channel we spotted a floating mat of kelp that had broken loose from the rocky shore of the mainland or one of the islands. Known locally as kelp paddies, these drifting masses of algae are fairly common in the Santa Barbara and Catalina Channels. Because the floating kelp gets plenty of sunlight and obtains its nutrients directly from the surrounding water, kelp paddies can thrive for months as they drift with the currents.

Many of the fish that live along the shore spawn by releasing thousands of floating eggs, which develop and drift with the plankton before hatching. Those larval fish that are fortunate enough to end up near shore at the right time of their development may survive, but many that drift helplessly in the vast, open sea either die or are eaten. Drifting kelp paddies serve as temporary oases of shelter for sometimes hundreds of young fishes that happen upon them as they drift.

Over time, the kelp paddy accumulates more and more young fish. Eventually, predators like yellowtail (*Seriola dorsalis*), tuna, and jack-mackerel (*Trachurus symmetricus*) begin to circle the paddies, trying to pick off the smaller fish that are using the kelp for shelter. Often, too, one can find a mola, or ocean sunfish (*Mola mola*), hanging below these floating islands of life, even though it isn't in search of food there.

Spotting this drifting kelp paddy, we knew it could be our chance to pick up the halfmoons we needed for our kelp gallery. Using little barbless hooks with colored yarn attached and casting next to the paddy, we caught a couple of dozen young three- to four-inch halfmoons. This was excellent; we hadn't even made it to the island yet, and we could cross the halfmoons off the want list.

The other species we were after required different techniques. We rendezvoused with the *Just Love* and prepared for our first dive, anchoring next to a kelp bed in a small, rocky-bottomed cove. Here we planned to collect two of the bright-orange garibaldi and a number of señoritas (*Oxyjulis californica*) and to scout around under ledges and boulders looking for sleeping horn sharks and swell sharks.

The trick to catching garibaldis and señoritas is to break open a red sea urchin to attract the ever-hungry señoritas. Before they can even smell the tasty urchin, they respond to the noise made when it's broken open with a knife. Resting motionless on the bottom next to the broken urchin, I waited with my hand net poised just above it. A half-dozen eager, competitive señoritas swam over, and a couple of them—with one eye watching me—moved onto the urchin. For a split second they looked down, intending to take a bite of urchin roe, and at that moment I swooped down with the net and they were caught. Quickly transferring them to the plastic bag tucked under my wet suit crotch strap, I was ready for the next ones. Attracted by the commotion, an adult garibaldi came over to check out the action. Keeping one eye on me, it moved in for a bite, but then it, too, looked down for one critical moment and into the bag it went. After ten minutes the plastic collecting bag was at capacity with eight or ten fish and needed to be taken up to the holding tank. After dropping the fish off, back down we went for more.

While two of us netted the señoritas and garibaldi, two other divers scouted for resting sharks. The horn sharks were easy to collect: all we had to do was pick them up and transfer them to our large net goody bags. Horn sharks are so named for the bony spine they have in front of each of their two dorsal fins. When the fish is newly emerged from the egg, these two spines are quite sharp, making for a painful surprise to any predator that tries to swallow the hatchling. As the shark grows, the spines become dull from rubbing on overhanging rocks; at that point, though, the shark's larger size is enough to discourage any would-be diners.

The defense system of swell sharks is quite different and can be a bit of a problem for a collector if one is found lodged under a low, overhanging rock. Their "swell" name is very appropriate: to defend themselves, they gulp water like a puffer fish and swell up to two or three times their normal girth. This tactic, along with their rough, sandpaper-like skin, allows them to effectively wedge themselves in place, making it very difficult to pull them out. I've never figured out what natural predator would try to pull a swell shark out from its resting place, but there must be one for such a complex and effective defense to have evolved. Still, it's not quite enough to deter us humans with our two skillful hands. Their defense tactic only serves to slow us down a little; we can work an inflated swell shark loose in just a few minutes. Then, holding it behind the head and by the tail, you pop it into your dive buddy's open net collecting bag. It is always wise to keep two hands on the shark; holding it by the tail only is an invitation to get bitten, since a flexible swell shark can whip around with its large mouth open and ready. Bagging a shark is no problem with two people, but it can get tricky when you try this by yourself. I generally got the head next to the bag's opening, then as quickly as possible let go, unlatched the bag, gave the shark a shove toward the opening, and hoped it went in. Most of the time this worked, but it was always exciting.

The next evening the *Just Love* and *Lucile* were anchored in a protected cove for the night. After supper we suited up and prepared for a night dive. The quarry this time was blacksmith, a damselfish relative of the garibaldi. During the day they are impossible to catch as they actively feed in the upper layers of the water column on tiny zooplankton. Like many other diurnal fishes, though, blacksmith spend the night resting in crevices in the rocky bottom, at which point they become easy prey.

This was the first night collecting the newer aquarists had attempted, and they learned by doing: first, how to catch the fish, and then how to transfer them from the net into the bag—all in the dark, and all while you're holding a dive light and a hand net containing an animal motivated by a strong desire to escape. At the same time, you're trying not to get tangled up in the kelp. Despite these challenges, within short order we had our quota of blacksmith and, low on air, headed back to the boat, eager to call it a night.

The next day both boats headed around to the seaward side of the island. We planned to collect some kelp bass with hook and line and, hopefully, to find a large male sheephead for our Kelp Forest exhibit. Working off the *Lucile,* we positioned the boat in the middle of a dense kelp bed and quickly caught several ten- to twelve-inch kelp bass. Bob Kiwala then decided to try for the sheephead.

One of the interesting aspects of the life of sheephead is that, like many other members of the wrasse family, they begin life as females. They have a harem system where a large male stakes out his territory. Within that territory, all the other sheephead will be females. However, if anything happens to the male, the largest female will change into a male. She will develop the hump on top of her head and change from a dusky pink to the striking red, white, and black colors of the male. Her ovaries will degenerate, and fully functional testes will begin to develop. This drastic change in body shape, color, and function can take place in as little as three or four weeks. As long as a large male is present, however, the other sheephead remain as females.

Sheephead, with their stout canine teeth, feed primarily on hard-shelled invertebrates like crustaceans and sea urchins. One of their favorite foods is lobster (which might be one reason lobsters come out only at night, when the sheephead are sleeping). Taking a stout rod, Bob baited a hook with a good-sized chunk of fresh lobster tail that he'd grabbed the night before in preparation for sheephead fishing. Dropping it down through an opening in the kelp canopy, he kept the bait a few feet off the bottom and waited.

All of a sudden the clicker on the reel sounded the alert: something had grabbed his bait and was making off with it. Putting the reel in gear, Bob pulled up hard and set the hook. He instantly knew he had something big. It was an even match for a while as the strong fish wrapped the line around a number of kelp plants. Bob slowly worked the fish, together with a wad of kelp, to the surface, and we finally saw what he had: a beautiful big male sheephead—just what we wanted.

Next we rendezvoused in a cove with the *Just Love.* Here Mark Shelley, the underwater photographer with the video production company Sea Studios, wanted to shoot some video to be used in an educational piece on collecting for the aquarium. We found a number of small female sheephead, which, like the señoritas, we baited in with sea urchins and then hand-netted. Swimming back to the boat after the video shoot, I spotted some small ocean whitefish cruising just above the sandy bottom of the cove. Back on board and out of our dive gear, we dropped small baited hooks down and caught several. Although not colorful, ocean whitefish have a delicate shape and graceful swimming motion and, we knew, would make a nice addition to the Kelp Forest exhibit.

By now our holding tanks were full, and it was time to take our first load of fish across the channel to Santa Barbara, then transfer them to the truck for the drive back to Monterey. With the objective of expanding their experience, a second team of new aquarists, under the guidance of Chuck Farwell, took over to collect the rest of the fish on the list. Our collection was taking shape and we were breathing easier as October 20, opening day, approached.

ANOTHER GREAT WHITE SHARK ENCOUNTER

Late one Sunday afternoon in September I received a phone call at home from marine biologist Paul Seri of the University of California's Bodega Marine Lab north of San Francisco. A halibut gillnet fisherman had brought in a young great white shark, he said; it looked active and healthy swimming around the lab's twelve-foot circular pool, and he wanted to know if we wanted it.

That was about the last thing I needed at that particular time. We still had serious work to do to get the aquarium open, and with the past experiences I'd had with white sharks at Sea World and Steinhart Aquarium, I didn't look forward to another heartbreaking failure. Silently hoping the shark would die, I told Paul Seri to see how it did overnight and to call me in the morning. The next day he said it still looked great and was cruising around the pool without any assistance.

Scott Nygren, who'd worked with sharks at Sea World, had been hired to fill the vacancy created by Mike Weekley's death. He and I loaded a transport tank, oxygen-injector pump, oxygen cylinder, and shark stretcher onto our truck and took off for Bodega Bay. Five hours later we pulled into the marine lab and were frankly surprised to find the shark doing very well, swimming around just fine in the small pool, as Paul had said. Lowering the water level and pulling on our hip boots, Scott and I climbed in the tank, guided the shark into the stretcher, and lifted it out and into the oxygenated tank on the truck. We took off immediately for Monterey.

Monterey Bay Aquarium's unexpected white shark, one month before the aquarium was to open. (Photo © 2000 Monterey Bay Aquarium Foundation. All rights reserved.)

Five hours later when we pulled in at the aquarium, the shark still looked good. Just as we'd done with our sevengills, we hoisted the transport tank up next to the 335,000-gallon Monterey Bay Habitats exhibit. The lid was removed, the tank was lowered into the exhibit, and the shark swam off. Slowly it began to swim around, avoiding the walls and window fairly well, but it listed to one side, presumably from being cramped for so long in the small transport tank.

After a while it straightened up and swam well, with strong, steady tail beats. The shark was definitely alert and was managing to stay away from the rocks and wharf pilings.

The next day we tried feeding it. We offered it a nice bloody, dead mackerel, but it showed no response, nor any interest in the many salmon, mackerel, or rockfishes swimming around it—all potential prey for a young white shark.

Over the next few days we tried on a number of different occasions to get it to feed. A live, struggling lingcod was dangled right in its path, but the shark repeatedly turned away and swam around it. All other attempts likewise failed. Mark Shelley came over from Sea Studios and took some excellent underwater video of the shark, and we continued with our regular maintenance duties, cleaning the windows of the exhibit and siphoning debris from the bottom.

I recall one day when I had siphoning duty. I was impressed by how alert the shark seemed as it cruised by, looking right at me. It was so different from the brain-dead white shark we'd had at Steinhart. This truly was a beautiful and awesome creature. In spite of its small size—less than five feet long—there was a presence about it that emanated power and grace. It gave me the same kind of feeling I have in the presence of a terrestrial equivalent, the Bengal tiger.

The white shark swimming so close by triggered very interesting behavior on the part of the four sevengills, all of which were larger than the white: although their normal swimming pattern utilized the entire exhibit, when the white shark was added, they moved to the deep end of the tank and stayed there. It was also strange that none of the smaller fishes, all potential food for a great white shark, showed any noticeable change in behavior.

On the eleventh day the white shark became disoriented, got stuck behind the wharf pilings, and died. Our conclusion was that, having refused all food, it had simply used up its limited energy reserves. Young great whites may have a higher metabolic rate than large adults and are probably not able to store as much energy. As a result, they may require more food more frequently. Warm blooded, like tuna and their relative the mako shark, they must continually lose body heat to the cold water. The young sharks have a higher surface-to-mass ratio than do the massive adults, which probably explains why they are normally born in southern California and Baja, where the water temperature is warmer. In normal-temperature years, only the large adults venture north into the cold water of central California, to feed on energy-rich elephant seals. Our shark was captured at the height of an unusually warm El Niño period, off Bodega Bay, far to the north of the normal range of a young great white. To maintain their internal temperature young sharks must need to eat frequently, but it's a mystery why ours refused all the food we offered it.

THE GREAT WHITE SHARK ENIGMA

The great white shark is one of the few top predators in this world that has little fear of being eaten. In that respect it's like the orca and tiger shark, or the big cats of the terrestrial world. Unlike these other top predators, however, the white shark has never been successfully kept in captivity.

Adult white sharks have been virtually impossible to capture and keep alive for a simple reason: their huge size. Juveniles of more manageable dimensions are only occasionally seen, and their capture has almost always been a matter of chance, typically occurring when a commercial fisherman finds a live one entangled in his net and calls an aquarium. Often by the time the aquarium staff have reached the capture site and then transported the shark back to the aquarium, it's in pretty bad shape. Because white sharks are ram ventilators and in the ocean must swim to pass water over their gills, their poor condition is usually the result of lack of oxygen during the time they were immobilized in the net.

In spite of these problems, young white sharks have occasionally been brought back to aquariums or oceanariums in good condition. They have looked good and behaved well, maneuvering actively and accurately around their tank, obviously alert to their surroundings. Based on criteria we use to judge other shark species, they should have survived and thrived. Yet despite their healthy appearance, in each case something was obviously wrong, because none of these sharks fed and all died, apparently from lack of food, after relatively few days.

I've had two white sharks that looked as if they would do fine. The first one, at Sea World, lived eight days, and the second, at the Monterey Bay Aquarium, lived eleven days. After I left Sea World, the staff there worked with three or four apparently healthy ones; one of these was force-fed mackerel and it lived sixteen days, the longest any young white shark has survived in an aquarium. At Steinhart Aquarium, John McCosker had one in the Fish Roundabout that initially appeared to be in good condition. However, it did not adapt well to the tank and after four days was taken out and released near the Farallon Islands. Finally, Ian Gordon at Underwater World in Sydney, Australia, captured one on a setline only forty-five minutes away from his aquarium. Unfortunately, the space in his aquarium was too confining for a free-swimming white shark, so after a few days he took the shark out and released it exactly where he'd collected it.

The failure of the great whites to show interest in food appears to be a result of stress during capture and transport, or it is possible that the aquarium itself alters their physiology, causing their normal appetite to shut down. It's ironic that the animal that was pictured in *Jaws* as being a ravenous eating machine refuses to eat anything at all in an aquarium.

It's puzzling, moreover, that this stress doesn't seem to affect other species of sharks. Predatory bull sharks and lemon sharks collected in much poorer condition, for example, have quickly and completely recovered and gone on to live for many years in an aquarium. The problem can't be a matter of size, because giant whale sharks thirteen to twenty feet in length are being kept at Okinawa and Osaka aquariums in Japan.

I believe that if the effects of this initial stress could be countered, white sharks would do well in an aquarium. A number of possible approaches exist, the simplest being an injection of a stress-reducing steroid or vitamin B_{12}. A more complex solution would be to intravenously counteract the changes that stress causes to the chemistry of the shark's blood, much as is done to a human suffering from shock.

A more holistic approach, and one that I had plans to use at one time, would be to net off a cove at an island. When a healthy shark was caught, it would be transported to the island pen and offered food. Once it began to feed (assuming it did), the netted-off area could be gradually reduced until it approximated the size of the aquarium that would eventually be the animal's home. The cove, thus restricted in size, would still have ample room for the shark to swim, but it would also have rocky walls and a shallowing bottom to adjust to. Perhaps someday we'll have an opportunity to give this technique a try.

The keeping of sharks in aquariums is an issue that has raised awareness of their value both as living creatures in their own right and as vital forces in maintaining balance in the natural world. Aquariums, as well as the many underwater films made of this graceful animal, have played a significant part in banning the deliberate killing of white sharks in Australia, South Africa, California, and Malta.

THE GRAND OPENING

By September 1984 everything was coming together. The artificial pilings and rocks that had been underwater in the Bay for two to four years were now in place, and the growth on them looked terrific. Everyone was busy with a long list of things to do, but somehow this busy schedule was rewarding. We knew we'd already accomplished a great deal, and our confidence was mounting: we were going to make it.

The exhibits were shaping up to be a good representation of the rich, diverse marine life of Monterey Bay, and I personally was pleased with them. Even so, I worried about whether the public would like the aquarium as much as we did. Would the average person—Joe from Kokomo or Ruth from Duluth, as David Packard referred to potential visitors—find it fascinating, or had we gone too far in trying to present a world that might be too alien? Would anybody even show up when we opened the doors?

We assumed some people would come because of the intense local interest in the aquarium, but how many? Estimates had been tossed around by our colleagues in the aquarium world, and they ranged from six hundred thousand to a high of one million for our first year. It wouldn't be long before we were going to find out.

Many of our projects required that several people work together, and a camaraderie developed among us. Even though we each had our own areas of responsibility, we helped one another out and felt we were all pulling together toward a common goal. It was a good feeling.

The marketing department had been busy too, and word of the new aquarium was spreading throughout California. Many of us did our bit with newspaper and magazine interviews and, of course, with the TV and film crews. The sea otter pups were the biggest hits with the media—proof, as if we needed any, that a high cuteness factor goes a long way. Everyone wanted to get the pups in their news pieces, and the poor overworked otter staff were swamped with requests for interviews, photographs, and video footage.

Because of earlier indecision in selecting an exhibit designer, the graphics production got a late start and delivery of them was running well behind completion of the rest of the exhibits. It was literally down to the last minute of the last night that the last graphic panels by the exhibit tanks were being assembled. Julie Packard, Linda Rhodes, exhibit designers Jody Armstrong and Jim Peterson, and writer Judy Rand were installing the graphics until midnight many a night before opening day. They kept going, I think, on pizzas and sodas, delivered to the aquarium at all hours.

Grand opening festivities had been arranged months in advance. Endless parties for VIPs and special groups were held during the two weeks before we officially opened to the paying public. On the big day itself, October 20, 1984, Cannery Row was closed to vehicular traffic. The response of Monterey Peninsula residents was very moving. Several local school and civic groups paraded down the legendary street, made famous in part by John Steinbeck and Ed "Doc" Ricketts, heading toward the aquarium. There were kids dressed up as all kinds of marine creatures—crabs, fish, octopuses, and a very imaginative class of little jellyfish. It was poignant to see the creative efforts these teachers and schoolchildren had made in preparation for this long-anticipated day.

The hour to open the doors finally came, and all the staff was there, as excited and full of anticipation as the anxiously waiting visitors. A sound system had been set up in front of the entrance, and Julie, Lucile, and David Packard gave brief but inspiring speeches. Then the first visitors were allowed to buy their tickets. The people loved the aquarium. Many stood awestruck in front of the tall Kelp Forest exhibit, with its gently swaying kelp and swarms of fish. They'd never seen anything like it.

In the days that followed people came in numbers beyond anyone's wildest dreams—certainly far beyond the highest predictions. The weeks, and especially the weekends, were incredibly busy, and the aquarium and Cannery Row were packed with visitors. The two most common complaints were the crowds and the lack of parking.

Only a little over a month before our opening, the white shark had been swimming around in the Monterey Bay Habitats exhibit. In a way, it's just as well it didn't survive. If we'd opened with a healthy great white shark on display, we would have had a logistical disaster on our hands. I could just imagine the traffic jam extending from San Francisco to Monterey as people flocked to see the first-ever live "Jaws" in a brand-new aquarium.

Many staff had worked extremely hard on this aquarium for a year to as many as six years. It was a good feeling to see the success of what we'd created, but at the same time I think we all felt a bit of a letdown after the opening—a sort of postpartum depression, if you will. The anticipation that had kept us going was all of a sudden gone, and there was nothing to take its place yet. Luckily, the feeling doesn't last long.

19

ALWAYS SOMETHING NEW

THE LIVE EXHIBITS AND THE day-to-day operation of the Monterey Bay Aquarium continued to improve during the years after opening. As expected, attendance dropped to a more manageable level, settling at around 1.7 million visitors a year, down from a hectic 2.3 million the first year. We began to think of ways to keep the aquarium in the public's eye and to give repeat visitors something new to see.

One section of the second floor of the aquarium had been left undeveloped for future use. The exhibits department came up with a series of marine-related art shows to utilize that space—photographs and landscape and seascape paintings, chiefly. One presentation, "Humpback to the Future: *Star Trek* at the Aquarium," documented the filming of part of *Star Trek IV: The Voyage Home* at the aquarium. In the completed film, computer-generated "virtual" humpback whales looked like they were swimming in our tanks. They seemed so real that some visitors who had seen the movie came to the aquarium expecting to see live whales.

These shows brought a few local people back, but they weren't a major attraction.

A larger exhibit called "Whalefest" followed in 1988. Scattered like a treasure hunt throughout the aquarium, this multipart exhibit featured the biology and behavior of the five species of giant whales found in Monterey Bay—always a favorite topic. But as with the *Star Trek* show, some visitors were disappointed because they didn't get to see real whales.

Meanwhile, the Monterey Bay Aquarium was doing well in every area—visitor satisfaction, encouraging ocean awareness, and earning revenue. Management began to consider the possibility of a major expansion, but while these discussions were going on, the aquarium decided to launch a series of live special exhibits, both to attract new and repeat visitors and to more effectively communicate our message of ocean conservation.

"MEXICO'S SECRET SEA"

One of the most memorable of these exhibits, at least to those of us in the husbandry department, was "Mexico's Secret Sea," which opened in 1988 and closed about a year later. This exhibit was the first of many in which the husbandry and exhibit departments worked together to present, in a fresh, stimulating way, live animals not necessarily found in Monterey. Since that time, in fact, this second-floor space has become the primary special exhibits space, focusing on a variety of topics, messages, and animals.

"Mexico's Secret Sea" didn't happen overnight—far from it. Planning began in 1986 when, having interpreted all the nearshore habitats of Monterey Bay, we decided to look beyond Monterey for a suitable topic. Steve Webster and I had each spent considerable time in Baja California, and we pushed hard for an exhibit showcasing the Sea of Cortez. John Steinbeck and Ed Ricketts's book *Sea of Cortez* provided an excellent connection between Cannery Row and Mexico's Gulf of California.

Finally, our proposal was accepted and an intriguing story line developed: the exhibit would retrace the scientific expedition that Steinbeck and Ricketts made from Monterey to the Sea of Cortez in 1940 aboard the fishing boat the *Western Flyer.* Although the husbandry staff would visit some of the same spots Steinbeck and Ricketts had seen, we would travel by land down the length of the Baja California peninsula and transport the live animals back to Monterey by truck and air.

My previous expeditions to the Gulf of California were of immense help in determining the live exhibits, selecting the species to be featured, and, eventually, collecting the animals. The husbandry staff and exhibit designers worked for a year and a half to design the exhibit. It also took a year to receive word, through the American embassy in Mexico City, that a collecting permit had been issued to us by the Mexican government.

Our plan was to collect our animals at Los Frailes, near the tip of Baja California. This was the same place where Bob Kiwala, Kelly McColloch, and I had tried to collect back in 1968, when our plans were disrupted by the unexpected storm. Chuck Farwell had made a collection there in 1978 for the Scripps Aquarium, and I'd driven down from Steinhart to join him. So I knew the place fairly well.

This time our exhibit would be relatively small, so we wouldn't need the large, floating receivers or the numbers of fish that Sea World had required for their large fifty-thousand-gallon exhibit. Nor would we need to transport the fish back to the States in holding tanks in a large boat. We planned to fly most of our animals out in Styrofoam boxes to San Francisco from the recently completed jet airport at San José del Cabo, not too many miles from our collecting site at Los Frailes—itself still accessible only by a dirt road. We would camp on the beach and anchor our small receivers just offshore. Although most of the animals we needed could be collected at Los Frailes, the returning trucks would stop on the way north at Bahía de la Concepción and Bahía de Los Angeles for finespotted jawfish, sea fans, and other invertebrates.

Chuck Farwell and the aquarium's new collector, John O'Sullivan (Bob Kiwala had left the aquarium to return to Rarotonga), worked out the logistics of the expedition and made seemingly endless lists of all the equipment we'd need. Not only did we have to deal with the usual problems of collecting, holding, and shipping the animals, but we had the additional challenge of living on the beach. Conditions would be very primitive—no electricity, no fresh water, no toilets—and we'd have to bring almost everything we needed with us.

Because of the intense sun, shade for protection from sunburn was at the top of our list. Two large tarpaulins lashed together and secured by guy lines to plywood sand anchors would be set up over the propane-stove kitchen and food storage areas. At night, each of us would be on our own to choose a comfy spot on the soft sandy beach to bed down and sleep under the stars.

A BAJA ADVENTURE

The time came for departure. The stakebed truck and pickup were piled high with equipment: ten scuba tanks, two air compressors (just in case one failed), two Zodiac inflatable boats and outboard motors, tables, chairs, oxygen, Styrofoam shipping boxes and plastic bags for sending animals back to the aquarium, not to mention personal belongings, kitchen equipment, and canned goods. Duplication was our motto: if any item was essential to our success, we took two. The large fish transport tank was filled with drinking water for the trip down. When it was empty, it would be filled with seawater and used to bring back the large morays as well as the animals collected on the drive back up the peninsula. The mountain of stuff we were bringing was mind-boggling.

The Mexican collecting permit we'd applied for a year earlier still hadn't arrived, but we had a letter from Mexico City stating that it had been issued, it was on its way, and we would receive it shortly. There were many reasons we couldn't postpone the departure date of the trip, and word that the permit was coming was good enough. The two trucks took off for Baja.

The plan was this: one group—consisting of John O'Sullivan (the lead man on this project from start to finish), aquarist Steve Brorsen, our former collector Bob Kiwala, and Ron McConnaughey, whom we had "borrowed," along with his Zodiac, from Scripps—would drive the two trucks the four hundred miles to San Diego, cross the border at Tijuana, and then continue straight on to Los Frailes, another eight hundred miles away. They would pick up four additional people flying in to San José del Cabo—aquarists Mark Ferguson and Carolyn Darrow, curator Chuck Farwell, and underwater photographer Flip Nicklin—and with their help get the camp set up. After several days of collecting, some of the early participants would fly back, and others would fly down to take their place. The third group would include aquarists Scott Nygren, John Christiansen, Bruce Upton, Dave Wrobel, Kelly McColloch, my friend from Sea World days, and me. Finally, a fourth group—aquarists Randy Hamilton and Gilbert Van Dykhuizen, videographer Chris Angelos, and diver and all-around mechanic Randy Wilder—would arrive, while others left. It would be the job of those who remained to break camp, make the two collections further up in the Gulf, and drive back to Monterey.

Although this plan was a bit of a human juggling act and would cost more than if we'd done it with just one team, I felt the experience and skills gained by all would be well worth it. In particular, I knew this arrangement would allow new husbandry staff to learn firsthand just what it takes to get an exotic fish from its native habitat to an exhibit, and thereby gain a real appreciation of the planning, hard work, and risk that goes into every step of the collecting process.

The two trucks made it all the way to Los Frailes without trouble and met the four fly-in aquarists at the airport. The next two days were hard, sweaty work as they set up the kitchen and the sunshade tarps in almost one-hundred-degree heat. Much cold beer was consumed, and frequent dashes made to cool off in the clear, eighty-four-degree ocean a few yards away. The two inflatable Zodiac boats were assembled and blown up with air from a scuba bottle. The two floating fish receivers were taken out and anchored a short distance off the beach in twenty-five feet of water, all ready for the first fish.

Los Frailes is a beautiful, wide sandy beach in a cove that's protected from most seas by a high, rocky headland. The site for our camp was near the mouth of a dry arroyo, or wash. Just inland from the beach the vegetation was typical semitropical desert, a mixture of cactus and low scrub trees, over which giant cardon cactuses stood like many-armed sentinels.

A group of six Mexican commercial fishermen were camped on the other side of the arroyo mouth. They fished from three twenty-two-foot-long fiberglass pangas with fifty-horsepower Johnson outboards, leaving every morning at sunup and returning in the afternoon. A panel truck filled with ice was parked at their camp; they drove it to La Paz every two or three days to sell the catch and pick up more ice.

The fishermen had developed a remarkable method of running their boats far enough up the steep beach that they would be safe from heavy

Our home on the beach at Los Frailes. (Photo courtesy John O'Sullivan)

surf that might develop during the night. After unloading the catch, the boat operator would circle out seaward, turn toward shore, and, giving full throttle, aim straight for the beach. At the last second, just before hitting the steep sand slope, he'd reach back and raise the engine. The speed and momentum carried the boat way up the beach, where it came to a stop. Unfortunately, launching the boats in the morning wasn't so easy and was done by all six fishermen dragging each boat down the slope and into the water. We enjoyed watching the fishermen—especially their daily high-speed beaching technique—and became friendly with them. On a number of occasions they gave us fresh *cabrilla* or *dorado,* the Spanish name for the Hawaiian mahi-mahi (*Coryphaena hippurus*), for our dinner.

When I arrived by air, John, Bob, Carolyn, and Steve had already made a number of collecting dives and had a good population of fish in the floating receivers. Circling closely around the outside of the receivers, using them as shelter from predators, were dozens of half-inch golden jacks (*Gnathanodon speciosus*). Young jacks of many species are often found in association with large objects, such as floating logs and seaweed, or even large sharks and turtles. One species of jack, the black and white–banded pilotfish, has made this association a career, spending its entire life in the close company of sharks. I think if we'd made a hole in the receiver the little fish would have gone inside, but then, of course, the fish inside would have come out. We needed golden jacks for our exhibit, and after chasing the speedy little fish round and round the receivers, we managed to catch some with our hand nets. In they then went with their future tank mates.

As I'd learned from my previous trips to the Gulf of California, each species requires a method of collection suited to its environment and its behavior. Some, like the goatfish, butterflyfish, and angelfish, are easier to collect at night; others, like the squirrelfish, during the day. The rainbow and sunset wrasses we collected during the day in my usual way, with sea urchin–baited glass jars. This was an excellent opportunity for the new people in our group to learn all of these techniques.

THE DETERMINATION OF A MORAY

We needed three big Panamic green morays (*Gymnothorax castaneus*) for our exhibit, and Gilbert and I had spotted a nice one living in a very shallow reef just off the beach. So one night, we went out to see if we could locate and catch it. The plan was to slowly waft some of the fish anesthetic quinaldine into the eel's lair until it became groggy. Then we'd reach in, pull it out, and slip it into our net bag.

We found the eel, and, using my hand, I started to waft the anesthetic into its hole. Things were going well, when all of a sudden the moray shot out of its hole right between us and hightailed it down the reef. We, of course, took off in hot pursuit, hoping it would duck into another hole and stop. We almost kept up with it, but the eel must have been a bit younger—certainly more athletic—than we were and we lost it. The fact that it was dark didn't help. All was not lost, though, because a few days later two other morays were successfully collected; these we put in an underwater cage for safekeeping until we were ready to head back to Monterey. Or so we thought.

Our evasive quarry, the Panamic green moray, Los Frailes. (Photo courtesy Robert S. Kiwala)

The next day when we went to check on them, the "escape-proof" cage, made from two-inch PVC pipe covered with strong netting, was empty. Clearly, we had grossly underestimated the moray's Houdini-like skill. Not only are morays extremely strong, but they also have the ability to push their tails between two objects to force them apart. We reasoned that they had done just that with our cage, slipping the end of their tail between the flexible plastic pipes and separating them just enough to squeeze their whole body backward through the gap.

Not one to be beaten by a fish, Bob Kiwala secured the holding box by lashing all the pipes together so tightly that nothing could possibly squeeze between them. One more eel was collected; this time the cage worked, and any plot it might have had to escape was foiled.

We still needed two more morays, but we hadn't seen any on our recent dives, either at night or during the day. Bob talked to a fisherman some distance down the beach from us who said he'd accidentally caught a couple of eels casting from shore. At this point we were beginning to worry that we might not find our morays by the time we had to leave. Anything was worth a try, so John and the other divers went to where the fisherman indicated and dove in daylight. They found it was loaded with morays, of all sizes. They actually caught so many that they could be picky about which to keep and which to release. Saving two nice ones and releasing the others, they took them down and placed them in the underwater escape-proof eel cage, and this time they stayed put.

All was going well, so John O'Sullivan decided to drive to La Paz to pick up our collecting permit. Little did he know what awaited him. Arriving at the office of the Secretaría de Desarrollo Urbano y Ecología (Secretariat of Urban Development and Ecology, or SEDUE for short), through which we'd originally applied for our permit, he found that the permit had in fact been issued by a competing government agency, the Oficina de Pesca, which oversees fisheries in Mexico. The SEDUE people said the Pesca permit wasn't valid and that SEDUE should have issued the permit.

Although John wasn't actually arrested, he was told not to leave town while they worked the matter out, an innocent victim caught in a jurisdictional battle between two competing Mexican agencies. Luckily, they understood his plight and issued him the second permit the next day, though not for the exact same list of animals as on the first permit.

The collecting had gone well, the shuttling of aquarists between Monterey and Los Cabos had been smooth, and nobody had gotten lost, hurt, or stranded. The time came to make the last air shipment of fish to San Francisco, break camp, pack all the equipment back onto the trucks, and head north to Bahía de la Concepción and Bahía de Los Angeles to complete the collecting. The big morays would ride in the large transport tank on the stakebed truck.

Breaking down camp and loading the two trucks went quicker than expected. The Mexican fishermen lent a hand, and for their help we gave them the propane stoves, the cookware, and the sun tarps, none of which we expected to need again for an aquarium project. I'm sure they had much more use for this equipment than we did in Monterey, in any case.

A finespotted jawfish pair, with the male carefully carrying his mouthful of precious eggs. (Photo courtesy Tokyo Sea Life Park, Rie Masho)

The first stop for the collectors was Bahía de la Concepción, where they hoped to collect spotted jawfish—a high priority because we were devoting an entire exhibit tank to them. The location was a bit of a gamble, since none of us had collected this species here before, but we knew the bay was well within the fish's range, so chances, we thought, were good.

Upon arrival, John, Randy Hamilton, Gilbert, and Randy Wilder swam out from shore with scuba gear and their hand nets and plastic bags. Each carried a short monofilament leader attached to a small, barbless, unbaited fishhook. After much swimming and searching, they finally located a number of jawfish burrows, with their characteristic ring of shells and pebbles around the entrance. Randy Wilder dropped his hook down the burrow the same way I'd collected blue-spotted jawfish many years before. When the fish grabbed the hook to toss it out of its burrow, it was a simple matter to set the hook and pull the jawfish out and into a plastic bag. In this manner, several spotted jawfish were soon collected.

With the jawfish checked off of our "want list" the team took off again, heading north for Steve Webster's place at Bahía de Los Angeles. Here they would dive for gorgonians and sea stars, which were destined for the exhibit of Sea of Cortez invertebrates.

Steve had given John detailed directions on where to find the animals we wanted. They unpacked, assembled, and pumped up one of the Zodiacs for a short run around to the outside of the bay. Gorgonians—spindly, feathery relatives of corals and sea anemones—are abundant in many places in the northern part of the Gulf; all the divers had to do was select the right sizes and species. Once back on shore, they tied cord around the bases of the gorgonians and suspended them in the second holding tank, on the pickup truck. That way the delicate animals wouldn't be injured by contact with each other or with the walls of the tank on the long ride home.

Much as the team would have liked to have stayed another day or two in this beautiful place, they were under pressure to get the animals safely back to Monterey. After making a water change in the moray tank, they reluctantly took off for Tijuana and the U.S. border. When they arrived at the border, the customs official took one look at the two trucks piled high with all manner of stuff and immediately sent them over to Secondary Inspection, where thorough checks are done.

Asked what they had in the big tank, the Monterey crew replied, "Eels." That was all it took. The customs officer told them to get out of there, sending both trucks on their way without looking at a thing. It certainly couldn't have been because of the aquarists' clean, innocent looks. Maybe he just didn't like eels.

The next stop was twenty miles north in San Diego, where the Sea World staff provided a much-needed change of seawater. Having replenished the holding tanks with nice clean seawater and anxious to get the fish to Monterey, our staff decided that with five drivers they could skip an overnight sleep stop in San Diego, and they took off in the early hours of the morning for Monterey. Ten hours later the aquarists pulled into the service yard of the aquarium. Tired and happy to be home, they each felt good about having pulled off a difficult yet highly successful collecting trip. The tequila bottle was passed around for a toast, and the five of them went home for a well-earned sleep in real beds.

There's no doubt that all of us in the husbandry department took special care of our new charges. We knew firsthand what it had taken to get those fish from underwater at the tip of Baja all the way back to the exhibits in Monterey, and we remembered each fish that we'd personally caught as if it were a longtime friend.

"Mexico's Secret Sea" was a great success. We knew that the animals, from the beautiful to the bizarre, would be popular in and of themselves, but Don Hughes and his team of designers did an excellent job of creating an exhibit that captured the atmosphere of Baja California.

In keeping with the Steinbeck-Ricketts theme, "Doc's lab" was reproduced, complete with authentic furniture, old books and specimen jars, and funky wooden aquariums containing live animals. Great attention to detail added that human touch, such as the pair of earrings in the ashtray by the bed. Throughout the exhibit were scenes of Baja California and of the collecting trip we'd just completed, contrasted with pictures of the 1940 trip that Steinbeck and Ricketts made on the *Western Flyer.*

"Mexico's Secret Sea," the first special exhibit that presented nonlocal marine animals to our visitors, proved a successful complement to our permanent exhibits. As it turned out, the expedition to Baja California was the last major exotic collecting trip the husbandry department would take, at least for a while. It was an unforgettable trip, definitely on a par with the one Steinbeck and Ricketts had made decades earlier.

I believe the significance of "Mexico's Secret Sea" was twofold: We learned that even though we were a regional aquarium, it was all right to have special exhibits that focused on animals from beyond Monterey Bay. We also learned that living exhibits are much more effective than nonliving ones at capturing the visitor's interest and getting our messages across. Beyond that, "Mexico's Secret Sea" literally opened the aquarium's doors to new approaches in the inspiring area of exhibition—approaches that would culminate over the next few years in a major aquarium expansion.

20

THE OPEN OCEAN

By 1987 THE MONTEREY BAY AQUARIUM was doing so well financially that the possibility of a major expansion began to be discussed. The aquarium's fiscal philosophy was modeled after David Packard's directive that from the day the aquarium opened its doors it must be self-supporting and avoid taking on debt. After three years of exceptionally high attendance and low operating costs the aquarium had accumulated considerable capital—enough, it seemed, to support major expansion. In addition, David and Lucile Packard had recently donated the old cannery next door to the aquarium for a possible expansion, having purchased it in the first place, apparently, to control future commercial development.

With the finances and site pretty much in hand, the next important issue had to be decided: What subject matter would the new wing focus on? Numerous exhibit themes were briefly considered during brainstorming sessions, and the aquarium eventually decided on the open ocean and the deep sea—the two largest and least-known habitats on earth—as the major topics. Not only would they make potentially exciting and innovative exhibits, but they would also complete the interpretation of the habitats of Monterey Bay.

Earlier in 1987 David Packard had founded the Monterey Bay

Aquarium Research Institute, devoted to research into the biology, geology, chemistry, and dynamics of the Monterey submarine canyon and development of the technology necessary to carry out this work. Among its considerable resources were a land-based research facility in Moss Landing, complete with laboratories, offices, and machine shops; an oceangoing research vessel; and a remotely operated vehicle, or ROV, capable of working to depths of about three thousand feet.

Although MBARI was funded separately from the aquarium, Packard's vision was that the two institutions would form a close and cooperative relationship that would benefit both education and science. The decision to interpret the deep sea as one of the two major themes of the aquarium's new wing was therefore a logical choice. In addition, we had been conducting valuable research at the aquarium itself—the progress we had already made in the keeping of jellyfish was a good start on the work we needed to do to develop open ocean exhibits.

I was well aware that, from the standpoint of exhibiting live animals, these two environments would be tough to pull off. There were so many unknowns. How, for example, would open ocean animals, living as they do in a vast world without walls, deal with the necessary boundaries of a tank?

The deep sea was even riskier. Just the collecting of fragile animals living hundreds or even thousands of feet below the surface was daunting. How well would they survive after being brought up from their cold realm of darkness, low oxygen, and high pressure? How would tanks need to be designed to accommodate these animals' different needs?

Pondering the big job before us, we decided the exhibits should be done in two separate phases: first the open ocean, with the deep sea coming some time later.

A great deal of R&D needed to be done on both subjects before we'd know what we could and couldn't do. Yet we also had an aquarium to run, and our top priorities were keeping the animals we had and maintaining the quality of the exhibits. The husbandry department staff was extremely busy.

Fortunately, the year before, I'd approached Julie Packard and asked if I could hire two additional husbandry staff, explaining the need to work on new ideas. In keeping with the Hewlett Packard philosophy

(continued on page 257)

THE YELLOW SUBMARINE

In the spring of 1991, Delta Oceanographics, developers of a small, two-person submersible, approached the aquarium and MBARI to see if we might be interested in chartering the vehicle for collecting and research purposes. Over the past several years the sub had made more than two thousand dives around the world, in, among other places, the Arctic, the Atlantic, and the Great Barrier Reef and at H-bomb sites in the South Pacific. They thought there might be some projects in the canyon where a live person on the bottom would be better than a remotely operated vehicle.

They trucked the sub up from Los Angeles to take some of us on dives in the Monterey Canyon. Gilbert Van Dykhuizen and I were invited along, and we jumped at the chance.

The little vehicle is the VW Beetle of submersibles. Small, uncomplicated, and reliable, it's fifteen feet long, weighs a little over two tons, and can dive to a depth of twelve hundred feet. Power comes from eight 12-volt batteries that run the lights and the golf-cart electric motor, which drives the propeller in the rear. Dave Salter, the pilot, sits upright on a chair in the center, the controls, life-support monitors, and sonar at his fingertips; his head is inside the cylindrical access tower, and he looks out of a series of portholes. The passenger lies prone between the pilot's feet and can look out several portholes in the bow. To descend, the submarine floods its ballast tanks with seawater; when the time comes to return to the surface, it blows those same tanks out with air supplied from two standard scuba bottles connected to a hose manifold.

The day I went out, two dives were scheduled, one for me and the other for Mary Yoklavich, a marine biologist working on the biology and population dynamics of deep-sea rockfishes. The little submarine had been loaded aboard MBARI's research vessel, the *Point Lobos.* We left at seven in the morning from MBARI's dock in Moss Landing and headed out several miles into Monterey Bay. Our destination: the Soquel arm of the Monterey submarine canyon, where we would dive to a depth of eleven hundred feet.

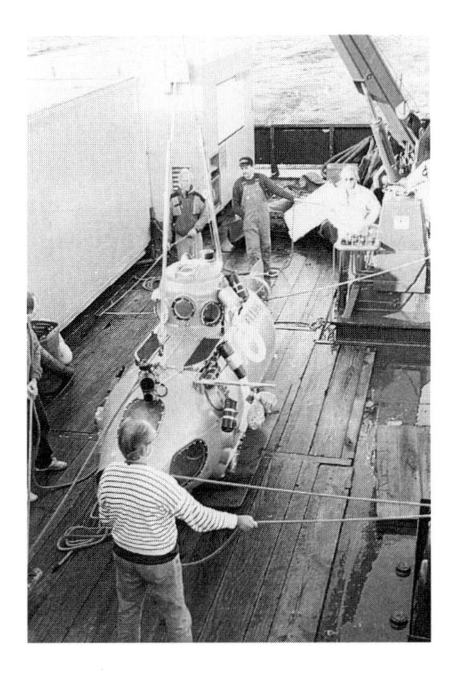

The little Delta sub preparing for a thousand-foot dive in the Monterey Canyon. (Photo © 2000 Monterey Bay Aquarium Foundation. All rights reserved.)

Arriving on-site, the *Point Lobos* slowed and scanned the bottom with its sonar until it was over the right spot. Dave described what to expect during the dive, including the safety features just in case anything went wrong down there. To offset its buoyancy, he explained, the air-filled sub has several hundred pounds of lead ballast attached to the bottom. If for some reason Dave were unable to blow the water from the ballast tanks, he could drop this lead ballast and the sub would rise rapidly to the surface. In case the propeller became entangled in a net or line and the sub became trapped on the bottom, the entire propeller section could be disconnected and the water blown from the tanks, and the sub would rise. There was also, he pointed out, enough oxygen and carbon dioxide–removal chemical to sustain life for about seventy-two hours.

We would be down for about two hours and toward the end of the dive, he said, it would get rather cold. The water temperature at eleven hundred feet is about forty-five degrees, and the steel hull of the sub has no insulation. He recommended wearing a warm jacket. Then, asking me what I weighed, Dave adjusted the ballast so the sub would be neutrally buoyant when on the bottom. I was ready to go.

With butterflies in my stomach, I climbed up onto the sub and squeezed through the access hatch on the top of the conning tower. It was definitely cozy inside; only a contortionist would be able to turn around. I lay down on a pad on the floor beneath the pilot's seat. Dave climbed in, closed and secured the pressure-proof hatch, and sat down, his feet on either side of me.

After Dave had given the okay to the hydraulic crane operator, we were lifted up, swung over the side of the ship, and lowered into the water. Clouds of silvery bubbles surged over the view ports as the sub rolled in the surface waves. This could be a seasick machine if you had to spend much time rocking and rolling on the surface. Throwing up in this tiny space wouldn't be fun for anyone. The crane hook was released, and Dave quickly powered us away from the side of the *Point Lobos.* Lying right next to a large steel ship in a rough sea is probably the most dangerous time of a dive.

Once we were clear, Dave asked if I was comfortable. I assured him I was okay, and we started down. The bubbles at the surface were gone and there was nothing but blue water ahead with countless particles of what is called marine snow (or, as marine biologists refer to it, "sea snot"). It's actually complex stuff made up of organic flocculent debris that originates in the sunlit surface waters—dead plankton particles, animal feces, mucus, and bacteria all clumped together—and slowly sinks. Scientists are finding that it serves an important role as a transporter of nutrients, the sun's energy, and carbon to the deep sea. Although I felt no sensation of descending, the particles of marine snow looked as if they were shooting up toward the surface. A fish or two shot by and also went up.

The blue became darker and darker until eventually it was completely black. As my eyes adapted to the dark I saw flashes and streaks of bioluminescence, the response of creatures disturbed by the passing sub. At one point during the thirty-minute descent, my mind wandered to the bizarre situation I was in, descending into total darkness under the pressure of hundreds of feet of seawater. I did a quick calculation and figured that when we reached the bottom there would be almost five hundred pounds of pressure on every square inch of the sub's surface. It was a sobering thought, but, recalling all the successful dives the sub had made, I quickly dismissed it and focused on what was happening in front of me.

Dave turned on the bright underwater lights, and a whole array of fascinating creatures became visible. There were red shrimps of several kinds; fish and an occasional squid shot by; and I recognized graceful sergestid shrimps, their long, slender antennae trailing behind them as they swam. Dave blew some air into the ballast tanks to slow our descent.

Then out of the gloom loomed the muddy bottom and, directly in front of me, a rocky outcrop covered with two-foot-high bright white plumose sea anemones. Hanging around the anemones were large vermilion and bocaccio rockfish. They were some of the biggest I had ever seen, but then, I had never fished or been down eleven hundred feet before. Nestled at the bottom of the rock were two or three red thornyheads, a species of rockfish found only in deep water. They're very difficult to collect alive, but we had a few small ones in the deep-sea R&D lab at the aquarium.

Dave cruised the submersible slowly over the bottom. A flatfish, possibly a species known as the deep sea sole (*Embassichthys bathybius*), swam away as we neared it. Here and there we saw hagfish (*Eptatretus stouti*) curled up in coils on the bottom. These remarkable eel-like fish are, together with the lampreys, the only living relatives of the ancient creatures that gave rise to the fishes and all the land vertebrates. Very primitive in physiology, they have no jaw, or vertebrae in their backbone; instead their flexible backbone is a simple cartilaginous rod. Almost blind, their rudimentary eyes can distinguish light and dark but cannot see images. Their most remarkable feature, however, is their ability to produce amazing amounts of sticky slime: a single hagfish can easily fill an entire bucket. This defense must work, because hagfish are rarely found in the stomachs of predators.

A common misconception among people is that if you drown at sea you'll be eaten by sharks. The truth is, you're much more likely to become lunch for a sucking, writhing mass of scavenging hagfish, which feed by boring inside the bodies of dead animals. Not a pleasant thought, to say the least.

Dave headed the sub into the slight current and gently settled onto the muddy bottom. Within a few minutes dozens of hagfish began showing up, attracted by the smell of the bag of fish pieces Dave had tied to the sub before we descended.

At the beginning of the dive Dave had shown me how the suction device worked. A stainless steel rod extended through a waterproof fitting to the outside of the hull; at its end was a flexible vacuum hose that connected to a pump and collection bag hanging from the rear of the sub. By turning on the pump, objects could be sucked up and deposited in the bag. By pushing and directing the end of the rod inside the sub, I was able to aim the hose at a hagfish; when I yelled, "Now!" Dave flicked on the pump, and up the hose and into the bag the hagfish went. We were able to collect several hagfish by this method—or more accurately, several hagfish and a lot of slime. The fish made it okay, but the net collecting bag was a total loss. Hagfish slime is almost impossible to get off.

We had been on the bottom for over an hour, and just as Dave had said, it was getting chilly. The walls inside the sub were wet from the moisture condensing from our breath, which trickled down to form a pool beneath the slightly raised platform I was lying on. I was growing increasingly uncomfortable, but there was no way I could change my position in the cramped quarters of the tiny sub. Although the whole experience had been incredible, when it came time came to head back to the surface I didn't complain: my aching body was ready. As we neared the surface Dave began to blow all the water out of the forward and aft ballast tanks. All of a sudden the sub started tipping at an alarming forty-five-degree angle, front down. I slid downward and ending up jammed, pretzel-like, in the bow.

Coming alongside, the *Point Lobos* got the crane hook onto the lifting eye, and we leveled off. Dave laughed somewhat sheepishly and explained that he had run out of air to blow out the tanks. He knew one of the two scuba bottles he'd loaded wasn't full, he said, but he'd thought it had enough air to do the job. All's well that ends well, I suppose, but considering where we had just been I felt it would have been nice to have a little extra air—just in case.

As exciting as it is to go to the bottom of the ocean, from the standpoint of collecting I believe the ROV is more practical, and it is cheaper to construct and to operate. Also, there are the safety factors and life support system that need to be designed for a live person in the sub. Even though chances are very slim that anything will go wrong, the fact that a living person is down there means that extreme care must be taken on every dive. On the other hand, the ROV is simply a highly sophisticated machine, and human lives would not be jeopardized if there should be an accident.

---

of dedicating resources to research and development, she said yes. The extra help would free up two aquarists from much of their daily routine, giving them time to solve the husbandry challenges inherent in our expansion plans—the first, and most crucial, step toward developing the new exhibits.

The two aquarists who would work on new animals were Gilbert Van Dykhuizen and Freya Sommer. Gilbert began research on the husbandry needs of deep-sea animals, and Freya on jellies. As it turned out, the work Gilbert and Freya did was exactly what we needed for the development and completion of the expansion, which would be called the Outer Bay Wing.

JELLY RESEARCH

The aquarium's jellyfish research actually started on a small scale in 1985 when Freya Sommer took on the challenge of determining whether jellies could be kept and exhibited year round.

Every organization is made up of individuals, each with her or his own personality, skills, and methods of working. Freya possessed an empathy for the animals plus an ability to focus her energies and approach her work problems scientifically. When given responsibility, the equipment and tools she needed, and support, and then left pretty much alone, she did fine work and helped us realize important goals. As Margaret Mead said, "Never doubt the power of a small group of committed people to change the world. That's about the only way it has ever happened in the past."

Freya's work with the jellies began with a list of known facts, from which she derived questions. For example, we knew that moon jellies (*Aurelia aurita*) are occasionally quite common in Monterey Bay, and for that reason they'd always been on our species list. Their occurrence is sporadic, however—sometimes they are here, sometimes not—which suggested that they had an unusual life cycle. This periodic absence from the Bay, combined with what was believed to be a life span of no more than a year, meant that we would have to figure out how to culture them—grow them ourselves—if we had any chance of exhibiting them. An ongoing culture would provide us with a steady supply for display during those times the wild jellies aren't available in Monterey Bay. And finally, we knew from their fragile nature that they would be a difficult species to keep, even if we did manage to raise them from scratch.

Most medusa-type jellies have an unusual, two-stage life cycle. The familiar swimming medusa, or "jellyfish," is either male or female and produces sperm or eggs that combine to produce thousands of microscopic larvae. If these larvae find a suitable surface to attach themselves to, they turn into tiny sea anemone–like polyps. These polyps are asexual and reproduce by repeatedly dividing, eventually becoming a colony of several hundred individuals. When conditions are just right, a polyp will change form and take on the appearance of a stack of little saucers. Each "saucer" (or ephyra) then splits off and swims away—a tiny medusa, only a millimeter in diameter, that will then grow up into a jellyfish.

This release of ephyrae is called strobilation. Strobilation is unpredictable, and jellies can remain in the polyp stage for long periods without producing the free-swimming form. This explains why some years jellyfish are everywhere and other years there are none. In any event, to provide us with a reliable supply of exhibit jellies, we needed our own colony of jelly-producing polyps.

Yoshitaka Abe, then curator of the Ueno Aquarium in Tokyo, had written two papers on the culture and exhibit of the moon jelly. We contacted Abe, and he kindly gave us some of his jelly polyps attached to a piece of plastic. The little polyps thrived, multiplied, and soon began producing tiny free-swimming jellies. With regular feeding, these grew and were soon large enough to be put on display.

Rather than try to reinvent the wheel, we decided to base our exhibit on Abe-san's plans. Using his design, aquarist Mark Ferguson built our first tank for moon jellies, the back of which consisted of a curved vinyl sheet perforated with hundreds of holes. The adult animals did well and were soon producing microscopic larvae by the thousands, which in turn settled out on surfaces in the jelly exhibit and developed into colonies of polyps. Freya used a razor blade to carefully transfer the polyps to sheets of plastic, where they were cultured in holding tanks and would hopefully provide a reliable supply of moon jellies for exhibit.

Although Abe's moon jelly tank was quite successful at keeping the jellies healthy, we thought its aesthetics could be improved. This particular challenge couldn't have been better timed. UCLA marine biologist William Hamner stopped in to see the success we had with the moon jelly exhibit and life cycle. He had designed and built a small tank for keeping planktonic animals on shipboard. It was modeled after an early German design called a "Planktonkreisel" that had been developed to study planktonic animals in a research laboratory. *Kreisel* is the German word for a child's spinning top—an appropriate name because of the circular water flow in the tank. Hamner said we could borrow his tank for a while and try it out.

Jellies are part of the plankton, which is made up, by definition, of organisms that are pretty much at the mercy of the currents. In conventional tanks with conventional seawater systems, plankton end up fatally stuck on the overflow screen or sucked into the pump. The secret to Bill Hamner's tank design was that the supply water flows smoothly across the outflow screen, gently blowing the animals safely away from

The delicate and graceful moon jellies. (Photo © 2000 Monterey Bay Aquarium Foundation. All rights reserved.)

the suction. Freya experimented with Hamner's tank and found that it worked splendidly for the fragile, delicately pulsing jellies.

John Christiansen drew up new plans based on Hamner's tank and we had one of our own fabricated out of clear acrylic. It, too, worked well and Freya began testing it with a number of different species of jellies. She found she could keep animals that before had been impossible to maintain.

AN INNOVATIVE DISPLAY TECHNIQUE

At this point it struck me that this basic design could be modified to create an exhibit tank that looked, so to speak, as if it wasn't there. The tank would have a rear-illuminated translucent blue back, to simulate the blue of the vast ocean, and we could take advantage of the angles of refraction of light passing between air and water to make the side, top, and bottom walls of the tank disappear. Like Bill Hamner's original tank, it would keep the fragile planktonic animals healthy and happy—if animals without a brain can experience "happiness." The end result would look like jellies swimming in the limitless blue of the open ocean.

To test this idea, John Christiansen and I added a thin sheet of translucent blue acrylic to the back of a transparent kreisel tank. We then reduced the size of the circular window opening so that the tank walls were out of sight. A fluorescent light behind the tank illuminated the blue back and a narrow spotlight shining in from the side of the tank lit the jellies.

Although the tank was rather small, the idea worked: all you saw were the illuminated jellies against an ocean-blue background. We also tested it on comb jellies (*Pleurobrachia* and *Beroe* spp.), and the side lighting made their iridescent rows of beating cilia ripple with rainbow colors. The comb jellies looked stunning, like shimmering alien space ships.

We knew then we were on the threshold of creating truly exciting jelly exhibits. Attempts to share our excitement with other staff, however, often proved frustrating. Some were impressed, but others saw only a small, improvised tank in a wet, funky lab and seemed unable to project beyond it to visualize how a future, large-scale jelly exhibit might look.

After her successes with the moon jelly, Freya became interested in the beautiful purple-striped jelly (*Pelagia colorata*), which in some years is quite common off the California coast. Virtually nothing was known of its life cycle, but biologists assumed that its reproduction was similar to that of its East Coast relative, *Pelagia noctiluca,* whose larvae skip the polyp stage and develop directly into young jellies within the female. Much to her surprise, Freya found that the California species didn't produce young swimming jellies at all, but, like the moon jelly, produced larvae that settled out and developed into tiny polyps, which in turn divided to form more polyps and some time later released tiny free-swimming jellies.

I remember an excited Freya bringing a bowl of the tiny jellies upstairs for Chuck Farwell and me to see. Not only was this an outstanding husbandry success, but it was also a scientific breakthrough. This discovery of the previously unknown life cycle of the West Coast *Pelagia* may eventually result in this jelly's being reclassified and placed in a different genus.

Next, Freya worked with the sea nettle (*Chrysaora fuscescens*), which at times occurs in huge swarms off the West Coast. Again she was successful in getting them to reproduce and produce polyps and then young medusas.

With all Freya's successes, we were running out of tanks and space. We designed and had larger kreisel tanks made to give her multiplying jellies room to grow.

Although the lives of planktonic creatures had been widely studied by a number of people in the academic world, few researchers had succeeded in keeping these animals alive for any length of time. Some of these scientists—most notably Claudia Mills of the University of Washington, Sid Tamm of Woods Hole Oceanographic Laboratory, and Ron Larson of Harbor Branch Oceanographic Institute in Florida—were excited about the work we were tackling and enthusiastically shared their knowledge with Freya. Their encouragement and support were of great help.

"PLANET OF THE JELLIES"

Freya Sommer's research results with jellies were first demonstrated in a special exhibit, "Living Treasures of the Pacific," which focused on the beauty of marine animals and plants: on their colors, shape, or motion as a form of living art. This exhibit, which opened in 1989 and ran for a year and a half, was then followed by "Sharks: They Are Not What You Think"—a topic we knew would be popular and easy to market. As we had anticipated, the exhibit was a success. Visitors learned about the diversity of sharks and that 95 percent of them are quite harmless, not at all like "Jaws."

Deciding on the topic for the next special exhibit wasn't as easy. Based on the progress we'd made in the culture of jellies, however, as well as visitors' reactions to the three "Living Treasures" jelly tanks, I pushed hard for an entire show on jellies. I knew full well it would be a hit.

In my experience, stimulation of the two sides of human experience—

the intellect and the emotions—is key to an exhibit's success. The emotional response we look for in designing an exhibit is usually pleasure, but the right brain can also be strongly stimulated by something gross or shocking. This could well explain the popularity in zoos of venomous snake exhibits, which touch on our basic emotion of fear. The left-brain orientation of an exhibit, in contrast, provides visitors with facts and information and appeals to their desire to learn. In the case of jellies, I knew that the beauty and grace of these gently pulsing animals would produce a strong emotional response in visitors, and their fascinating life cycles and anatomy would satisfy visitors' intellectual needs.

The marketing department wasn't at all thrilled about the idea. It's one thing to promote an exhibit like "Sharks," where all you have to do is mention the word "shark" and people will come, but the word "jellyfish" simply didn't have the same appeal. If it did anything at all, it conjured up a negative impression of something that stings or a disgusting-looking blob on the beach. It certainly didn't sound like something you'd drive a hundred miles to see.

Nevertheless, the aquarium decided to go with jellies as the next special exhibit, realizing it would be a challenge for the marketing staff. We aquarists knew that once visitors saw the animals, they would be enthralled; it was getting them to come that would be the challenge.

The title "Planet of the Jellies" was chosen; it played on the idea that jellies are like alien creatures, but at the same time it hinted at their true origin: our planet, the Water Planet. The story line was that of a voyage, with oversized guidebooks explaining the strange inhabitants visitors encountered as they made their journey through the exhibit. And the exhibit designers had a field day with the theme, creating flowing, curved walls that ran throughout the exhibit, giant jelly models that hung from above, and specially composed music, which added to the ethereal feeling of the space.

In the two years prior to the opening of the exhibit we worked with several species of tropical jellies. The upside-down jelly (*Cassiopea xamachana*), with a symbiotic alga living within its tissues, spends most of its time lying upside down on the bottom, photosynthesizing in the sunlight. A graceful Japanese species, *Tima formosa,* was given to us by the Enoshima Aquarium in Japan. The white-and-lemon-yellow, polka-dotted *Mastigias papua,* from the islands of Palau, added a touch of humor with its clownish activity. Several varieties of these jellies can be found in Palau, some of which have become trapped in small, landlocked saltwater lakes known as the "Jellyfish Lakes."

All in all we had twelve different species of jellies, ranging in size from the large and beautiful, golden-colored sea nettle to a little freshwater jelly that we collected from a vineyard irrigation pond in Napa Valley. All did quite well, though we had problems at first with the Palau jellies, which didn't seem to be growing as they should. We were fortunate to be visited early on by Kazuko Shimura, the jelly expert from Enoshima Aquarium. She had worked with more species of jellies than anyone in the world, and, despite the language barrier, she managed to point out that we needed more light for the Palau jellies to photosynthesize. Taking her advice we doubled their light, and their growth took off.

Partway through the show an aquarium member brought us a jar containing a small medusa that, she said, had appeared that summer in large numbers in the Petaluma River above San Francisco. My daughters and I used to water-ski in that river when we lived in Marin County, so, being a believer in having fun at work, I hitched up my ski boat (and the skis!), and Freya, her friend Peter Ferrante, and I took off for Petaluma to find the mystery jellies, and maybe get a little skiing in as long as we were there.

We found the little walnut-sized jellies in the boat-turning basin right in downtown Petaluma. It turned out that they are a brackish-water species from the Sea of Azov in Russia, probably transported here in the ballast tanks of a ship. Their introduction to the region, unfortunately, may have serious impacts on local marine life—especially the tiny fry of the striped bass population of the brackish San Francisco Bay and Sacramento Delta—because of their effective feeding strategy. Rhythmically pulsing their bell, these jellies swim to the surface, stop, and then spread their tentacles out like a net and drift slowly down to the bottom. Any little creature in the way of this stinging net is captured.

Of course, there's no way to get rid of the jellies now, even if they do start decimating fish. Once introduced into California waters, they're here to stay, like several other alien creatures that have shown up in

San Francisco Bay. To inform visitors of this ecological problem, one that exists worldwide as transportation becomes easier and farther-reaching, we set up a special display of these Russian jellies and gave it the eye-catching title of "Alien Invaders!"

"Planet of the Jellies" was an overwhelming success—the most innovative and also the most popular of the several special exhibits the aquarium has done, up to then or since. Even marketing the idea of jellyfish was successful as people told friends that they just had to go see this fascinating exhibit. John Racanelli, the aquarium's marketing director, summed it up when he said, "I thought we were crazy to put jellyfish on a highway billboard, but it turns out our visitors tell us that jellyfish are on their list of top three most popular animals."

The special exhibits we created, all featuring live animals, have clearly been the secret to the continued success of the Monterey Bay Aquarium. The repeat attendance that these exhibits generated kept the annual attendance at 1.6 million visitors or higher. I believe that without those special exhibits our attendance would have gradually declined, just as it has at other aquariums that lack new attractions to bring visitors back.

A LOOK INTO THE DEEP

At the same time we began work on the husbandry of open sea animals, we started considering the feasibility of keeping deep-sea animals. The oceans occupy 70 percent of the earth's surface, but the deep sea, unlike the two-dimensional surface of the land, is a three-dimensional space. If the deep sea is considered to begin below the photic zone—the level penetrated by usable sunlight—then the deep sea makes up over 90 percent of the earth's biosphere or inhabitable space. Very little of this vast volume has been explored. The deep sea is truly the last frontier on earth, and the strange animals that live there are far more of a mystery than those of the upper sunlit waters of the oceans. Yet if we approached this challenge methodically, we felt, we should have some success.

Acknowledging our ignorance about this undertaking, we began by inviting a group of deep-sea research scientists to participate in a workshop. We wanted to learn from them as much as we could about the biology and habits of the particular animals they studied, and find out whether they thought it was possible to keep these animals alive. Knowing which ones would most likely do well and which wouldn't could, we knew, save us considerable time.

Although few of the researchers had attempted to keep these animals for any length of time, they were most helpful and were optimistic about what we were doing. We had two major advantages at the outset: first, we were close to the largest submarine canyon in North America, and second, we had the cooperation of MBARI, whose research work focused on the depths of the canyon. Encouraged by the information our consultants gave us, and by their optimism, we started work.

Gilbert Van Dykhuizen launched the research into keeping deep-sea animals by setting up a series of holding tanks connected to two separate life-support systems and refrigerated to about 42°F, the temperature of the water at a depth of three thousand feet. For these initial research efforts, the water systems would be at surface pressure, with no attempt made to keep animals under high pressure.

In cooperation with MBARI and California State University's Moss Landing Marine Laboratory, we began collecting animals from the depths of the Monterey submarine canyon using MBARI's remotely operated submersible as well as midwater and bottom trawls and traps. Although deep-sea animals had been collected previously, they almost always came up dead or dying from damage caused to their fragile bodies by the nets used to catch them. The ROV, however, has special devices that were designed specifically to collect animals without damaging them: a maneuverable manipulator arm that can pick up rocks with attached animals, a controllable suction device, a storage drawer, and closable sampling cylinders. These proved invaluable for collecting extremely fragile midwater animals.

As in any pioneering venture, we had some successes and we had some failures, but what mattered was that we were learning more and more about the lives of the remarkable inhabitants of that deep, dark world.

Among our earliest successes were some fascinating animals. Towing a midwater trawl at depths of two thousand feet, we collected and have now kept for several years a number of mature eighteen-inch-long filetail catsharks (*Parmaturus xaniurus*). Like their feline namesakes, they have green reflective eyes for seeing in almost total darkness. We suspect they use their sensitive eyesight to find bioluminescent prey.

Searching the canyon walls at depths of around fifteen hundred feet and using the manipulator arm of the ROV, Gilbert worked with MBARI scientists to pick up rocks with soft corals attached to them. These strikingly beautiful mushroom soft corals (*Anthomastus ritteri*), with clusters of eight-tentacled polyps covering their globe-shaped body, did very well in our refrigerated aquariums, despite their delicate appearance, and thrived on blended krill shakes and brine shrimp.

One of the strangest creatures we collected was a predatory tunicate (*Megalodicopia nians*). Its nearest relatives are shallow-water attached animals that make their living by filtering plankton from the water. Rather than filter-feeding, this tunicate, which lives attached to the rocky bottom of the canyon floor, lies in wait for small prey to enter its open hood, to be captured in a manner similar to that of the carnivorous plant the Venus flytrap. Unfortunately, although these tunicates did well for a few months in our refrigerated tanks, they then declined and died. Gilbert set about figuring out why.

One significant feature of the deep sea off California is a zone between 1,600 and 2,600 feet, known as the oxygen-minimum layer, where the oxygen concentration is extremely low: as little as one-tenth of that found near the surface. We suspected that this tunicate was adapted to living under these very low concentrations, and our system, being exposed to atmospheric air and 100 percent saturated with oxygen, was actually toxic to it.

Gilbert therefore set up a special life-support system with low oxygen. It stripped oxygen out of the water by bubbling nitrogen through a small contact tower connected to the refrigerated recirculating water system. The water surface of the tank was covered with a plastic sheet to minimize the absorption of oxygen from the air. This oxygen-stripping system worked well, and Gilbert began to have success with the predatory tunicate. They were now staying alive and healthy for over two years, which was probably their normal life span.

In addition to these successes, however, there have been failures. We've had almost no luck, for example, with the delicate hatchetfishes and lanternfishes or the cephalopods that inhabit the midwater zone—that space that lies well above the bottom but below the photic zone, the level penetrated by light from the surface. Within the midwater zone, we have managed to collect and keep animals such as eelpouts (*Melanostigma pammelas* and *Lycodapus mandibularis*), slender, three- to four-inch fish with pronounced lips that look like they're pouting, but other desired animals have proved elusive.

As encouraged as we were with our very real successes, we found it sobering to think that thousands of unexplored feet of water lie below the depth—down to about three thousand feet—we had been working in. The prospect of keeping animals from those great depths alive are extremely slim, if possible at all. The animals that live there require high pressure for the functioning of their physiology—in short, for life itself.

In the end, although we couldn't have it all, the pioneering R&D work accomplished by Gilbert and the MBARI crew was key to the aquarium's ability to display deep-sea animals at all. Without their solid research, our attempts to present the new frontier of the open ocean and the deep sea would never have succeeded.

21

PELAGIC FISHES

In planning the aquarium's open ocean expansion, the first step was to come up with a species list of potential exhibit animals. We knew we would do major exhibits on jellies. Another main goal was to have a huge tank that would exhibit a community of pelagic, or open sea, fishes—a group that had rarely been displayed in public aquariums.

The idea was to represent the open ocean environment with as many species as possible that could get along together in one tank. Thus, not only did we have the challenge of collecting and keeping these fishes, but we had the problem of compatibility as well. Even beyond the possibility of them eating one another, would they all be able to get the food they needed, or would the more laid-back fish be outcompeted by the more aggressive feeders? There was a lot to learn, and there was only one way to do that: we just had to give it a try.

Open sea fishes tend to be migrators, visiting Monterey only seasonally or during El Niño years, when warm water moves up from the south. Schools of albacore tuna, for example, come by each year on their migration route north to Canada and across the Pacific Ocean to Japan. The bizarre-looking ocean sunfish show up in Monterey Bay almost every year during the warmer-water times of summer and fall.

Fishes like bonito and California barracuda from southern California are seen in Monterey occasionally, but then only during strong El Niño years.

Given the sporadic occurrence in Monterey Bay of many of these fishes, I knew we had to collect them where and when they were most likely to be abundant. Once they were in the aquarium, keeping them healthy and strong in captivity was another hurdle. Albacore had never been kept in any aquarium. In the 1960s, Marineland of the Pacific had kept bonito and barracuda quite well, but the aquarium staff collected them at their doorstep, so to speak, and didn't have to deal with the added problem of transporting the fish over long distances. We had our work cut out for us, there was no doubt about it.

OCEAN SUNFISH

Our first challenge was to work with the ocean sunfish. This animal—whose scientific name, *Mola mola,* means "millstone" in Latin, no doubt because of its flat, round profile—must be one of the most eccentric-looking but delightful fish in the world. Their rather silly open mouth and expressive eyes plus their strange, lumpy shape make them most appealing to anyone who sees them. Indeed, almost nothing about them resembles a "normal" fish. Their tail has practically disappeared, leaving a stumpy appendage that functions as a rudder, and they swim by sculling motions of their tall dorsal and anal fins. A thick, sticky slime covers their skin, beneath which is a layer of cartilage that is almost like a subcutaneous skeleton. They feed primarily on jellies of various kinds, yet despite this nutritionally poor diet they grow to enormous size, measuring as much as ten or twelve feet in diameter and weighing up to 3,800 pounds.

No U.S. aquarium had had much success keeping molas, but some Japanese aquariums had displayed them for a number of years by themselves in relatively small tanks. To prevent the molas from rubbing on the wall, the Japanese aquarists lined their tanks with a transparent vinyl curtain suspended a few inches from the wall. I found it distressing to see these large fish cooped up in such small spaces. My hope was that in a very large exhibit tank such as the one we had in mind

John Christiansen feeds an eager and cooperative ocean sunfish. (Photo

they would have plenty of room to swim around and could easily avoid the walls.

In late summer and fall, when our warmest water temperatures usually occur, small molas—less than two feet long and weighing twenty to fifty pounds—show up in Monterey Bay. We have no idea where they come from or why they come here. Whatever the reason, it would probably be a lot healthier for them if they didn't, since they invariably fall prey to the hundreds of California sea lions that are also here at that time of year. The agile pinnipeds catch the young molas and then fling them about at the surface like frisbees, presumably trying to break them up so they can eat parts of them. It is, at best, an inefficient technique; we often see dead or dying molas on the bottom of the Bay with their fins torn off by sea lions. It's very sad to see a helpless mola lying on its side, still breathing but unable to swim. Some years hundreds of dead, undamaged molas have been seen on the bottom. Although no one knows for sure, the assumption is that they come into the Bay with a current of warm water, then get caught when the temperature takes a sudden drop.

The first step of our mola research was to figure out how to catch and transport one. Feeding as they do on jellies, it seemed impractical to try and catch one on a line. (How would we get the jellyfish on the hook to begin with?) Molas are, however, often easy to approach by boat; once I'd even caught one by just leaping off a boat into the water and grabbing it. Although I didn't think that would be practical for our purposes either, it did mean the fish might be easy to scoop up in a net from the bow of a boat.

Finally we heard reports that molas had been spotted in the Bay, and our collectors prepared to go looking for them. Their dorsal fin can often be spotted at the surface. At first glance you might think it's a shark fin, but the shape and the characteristic back-and-forth sculling motion is quite different from the straight-line motion of the fin of a cruising shark.

One calm, flat morning, a fin was spotted, and the scooping method from the front of the boat worked. (The collectors learned they had to be quick, though, before the mola woke up from its reverie and took off when it realized a boat was bearing down on it.) Once caught, the mola was transferred to the small holding tank on the boat, brought to the aquarium, and released directly into the Monterey Bay Habitats exhibit. It looked great as it cruised slowly around the tank, and the visitors responded immediately to this strange new addition, certainly the most eye-catching of all the creatures in the exhibit.

Our next problem was to figure out how to get food to it in a tank full of other hungry fish. We knew molas eat jellies in the wild, but we also knew that providing them with a year-round supply of those animals would be impossible. Our hope was that we could switch the mola over to some other type of nutritious food that we could provide every day. For the moment, however, Freya Sommer offered some of her extra moon jellies as mola food. A few were released into the tank, and the mola almost immediately started to suck them up.

Until now Freya hadn't shown much interest in fishes, being fascinated instead by the diverse anatomy and life strategies of the so-called lower animals, the invertebrates. Perhaps she thought fish were mere evolutionary upstarts that hadn't yet been around long enough to deserve her respect. That first mola, however, intrigued her. After all, another creature that liked jellies couldn't be all that bad.

Together John Christiansen and Freya took on the challenge of caring for our new mola. Freya's extra jelly supply quickly ran out, so they tried feeding the mola small pieces of shelled fresh shrimp attached to a clip on the end of a pole. The fish loved it. Within a day it learned to associate the white plastic pole with food and would hurry over as soon as it spotted it. In spite of all the competition for food in that exhibit, the mola was doing very well.

After a few weeks, however, the temperature in the Bay dropped to fifty-three degrees. The mola stopped feeding and became disoriented. Sadly, we had just discovered the lower temperature threshold of molas. Because we had no way of heating that exhibit, the mola died.

We had learned an important lesson, though: if we wanted to continue research with molas, we had to have a temperature-controlled environment. We therefore installed a gas-fired pool heater, a recirculating pump, and a sand filter on a twenty-foot-diameter tank in the quarantine room. To accommodate the vertical nature of molas, we raised the height of the tank from four to six feet. We were now ready to try again.

Molas were still around, and John Christiansen and Freya collected four more molas for the new temperature-controlled tank. They found that the molas transported better when they were placed head down in a large plastic trash bucket; with no room to move, the fish couldn't damage the tips of their long fins. The trip in the bucket was quite short, and they were back at the aquarium within thirty minutes. Freya and John also designed a special stretcher for moving the oddly shaped molas from the boat to their new home.

One of the four new molas died within the first few weeks, and an autopsy showed it was loaded with internal parasites—large tapeworms in the intestines and nematode roundworms throughout the muscles—which, together with the stress of being collected and the strange environment, had caused it to decline and die.

We knew that molas are notorious for being heavily parasitized, and we decided to try to rid them of these unwanted hitchhikers. Their bodies were crawling with copepods and parasitic flatworms or flukes, but these external parasites were relatively easily removed with a treatment of formalin. The internal parasites, however, were a more serious problem. Our veterinarian, Tom Williams, suggested a drug called Praziquantil, which is used to deworm cats and dogs. The molas were all feeding readily from our hands, and it was a simple matter to hide a small pill in a morsel of their fresh shrimp. The next day we were delighted—and a little disgusted—to see balls of tapeworms lying on the bottom of their tank. Judging from the size of the balls, some of those worms must have been fifteen feet long.

The molas were doing well and growing rapidly until about two months later, when we began to have trouble with the water system and water quality. The rapid growth of the fish, which led to greater food intake, simply overloaded the water treatment system, resulting in a buildup of waste products beyond what the system could handle. Two of the molas went off their food, began to deteriorate, and died. Autopsies revealed that the deworming treatments had worked and they were completely clean of parasites. Meanwhile, the one remaining mola thrived and kept growing at an alarming rate.

Freya kept track of exactly how much food each mola was eating each day, and it seemed as if they were noticeably larger each week. After fourteen months, the remaining mola was so big that when its dorsal fin was at the surface, its lower anal fin was close to touching the bottom of the six-foot-deep tank.

Because of its size, plus the fact that we needed the tank for our bonito and barracuda research, we decided to release the mola while it was still small enough to be safely moved. The warmer water temperature in the Bay also influenced the timing of this operation. John O'Sullivan made up a four-foot-diameter stainless steel ring inside which he laced a circle of flexible, smooth plastic (it looked something like a giant pizza pan), and this we slipped beneath the mola. The giant fish was laid flat on the plastic, and with sheer muscle power and considerable effort we lifted it up and over the edge of its tank and into the transport tank, which was then lifted by crane into the collecting boat on its trailer. Once safely in the boat it was taken out to sea and released. Although we didn't weigh the mola before its release, we estimated it to be around three hundred pounds. In just fourteen months it had gained some two hundred and fifty pounds! That had to be one of the fastest-growing fish in the world.

We began to have rather fearful visions of how immense a mola might grow in a large exhibit tank with an unlimited supply of food. How would we deal with such a large fish if it ever had to be moved?

In 1998 I found out for myself. On August 7, 1997, a young mola weighing fifty-seven pounds was collected from Monterey Bay. It was treated for parasites in quarantine, where it quickly learned to take food from the hand. A month later it was moved to the new, million-gallon Outer Bay exhibit, where its primary caretaker, Tim Cooke, fed it shrimp, squid, and a nutritious gelatin-based blended food containing fish fillet, squid, prawns, multivitamins, and amino acids. Like our previous molas, it too had a phenomenal growth rate. After a year the staff realized that it would have to be released, and soon, before it became too big to fit through the channel gate to the outside holding pool.

By now it was six feet long, weighed an estimated six hundred pounds, and was far too large to fit in the holding tank on the *Lucile.* An air lift seemed the only way. John O'Sullivan constructed a large, strong, frisbee-shaped lifting device that would support the mola lying flat on its side. After obtaining permission from the Federal Aeronautics Board, he then arranged for a helicopter to fly over the aquarium and pick the mola up at dawn on November 4, 1998.

Everything went smoothly. Lured with food, the mola swam into the holding pool, where it was quickly weighed. It then took only thirty seconds for the helicopter to lift it from the pool and whisk it out over the Bay to a waiting boat. The mola was lowered into the water, where divers disconnected one side of the lifting bridle and watched as the fish swam off into the very sea it had come from fifteen months earlier.

As it turned out, we had greatly underestimated the weight of the mola. Instead of the 600 pounds we'd guessed, it weighed a whopping 880 pounds—a far cry indeed from its original 57. Undoubtedly, if we hadn't released it, it would have kept right on growing too.

BONITO AND BARRACUDA

Our next R&D project was to collect bonito and barracuda in southern California to see if we could successfully transport them three hundred miles north to Monterey. The marine science staff of the University of California at Santa Barbara helped by allowing us to set up a plastic swimming pool to hold our fish prior to transporting them.

The fish showed up off Santa Barbara in the late summer of 1989. Working from the *Lucile,* collector John O'Sullivan and his crew trolled small, shiny lures with barbless hooks and caught a number of both bonito and barracuda. Just as we'd done with salmon, the collectors carefully lifted them over the side of the boat, taking care not to touch them, and quickly released them into the onboard holding tank. Within a few days enough fish had been caught.

After a few more days in the temporary holding pool, it was time to transfer the fish to the truck and drive them to Monterey. John, attempting to move the barracuda with the same method we'd used with salmon—by lowering the water level, climbing into the tank, and guiding the fish into plastic bags—soon realized that another method was needed. Instead of swimming easily into the bag, the fish took off like pointy-nosed three-foot-long missiles right through the end of the bag, as if it wasn't even there.

John and his crew also learned that transporting barracuda and bonito together in the same tank is not a good idea. The problem lay in differences in behavior. Bonito, like all their tuna relatives, need to swim constantly in order to get oxygen, but barracuda don't: they can just hang motionless in the water. This meant they kept getting in the way, throwing off the steady swimming pattern of the bonito, which often gave up and sank to the bottom of the tank.

We learned from those initial mistakes, however, and soon the capture and transport of these fish was successful. Both species settled into their quarantine holding tank at Monterey, began feeding—the active bonito almost immediately, followed a couple of weeks later by the more cautious barracuda—and did well. Another hurdle was out of the way. Having actually done it, we now knew that we would be able to collect sufficient numbers of both of these fish for the Outer Bay exhibit.

We knew that if we could pull it off, the open sea exhibit would be unique, literally one of a kind in the world. Of course, collecting and keeping pelagic animals was unexplored territory for us, but we already had a good start on that critical part of the project and felt confident about getting a representative group of fishes to display.

However, one important challenge remained to be tackled if this exhibit was going to give our visitors the true feel of the open ocean. Unlike the other habitats we'd done in the aquarium, the open sea is nothing but water, with no visible objects other than the living inhabitants. Obviously, we had to have walls to contain the water of the tanks, yet we wanted people to have the impression they were looking into nothing but endless ocean. As much as possible, we wanted them to feel as if they were part of that environment. Somehow, therefore, the tank walls would have to be out of sight, or at least as inconspicuous as possible.

We had solved that problem in "Planet of the Jellies" with tanks that had translucent blue backs, creating the illusion of limitless water. We knew we could do the same with the new jelly displays, but the giant pelagic-fish exhibit we had in mind had to be made of concrete, and there's no such thing as translucent blue concrete. Someone suggested that an acrylic tunnel going through the tank would give visitors a sense of immersion. Yes, but the drawback of such a tunnel is that it's impossible to hide the surrounding walls of the tank.

I wrestled with the problem of how to create the impression of vastness in a limited space, and as an idea popped into my head I would jot down a sketch on an envelope, cocktail napkin, or whatever happened to be handy. The problem was not just to make the walls of the tank disappear, but to do it without making the tank so huge that we couldn't afford to build, operate, or provide space for it. Eventually I narrowed the possibilities down to one design. The tank would have to be quite large, but not unreasonably so. Its end walls would be totally out of sight thanks to the physics of light refraction. The back and the bottom of the tank would be curved like the inside of an egg, with no sharp angles that would catch the eye. And the whole thing would be blue, with the color shading from light blue near the surface to deep blue near the bottom.

There was to be viewing from two levels. To prevent visitors downstairs from seeing the tank bottom not far below their feet, I slanted the top of the lower viewing window back toward the viewers. This window also served to direct the visitor's attention up to the circling fish above.

Finally, if we couldn't actually re-create the open sea, at least we could give visitors an inkling of its immensity, and a huge window in the upstairs viewing area seemed just the ticket. Great advances had been made in recent years in the manufacture and installation of very large panels of strong, perfectly clear acrylic. We enlisted the talents of the Nippura Company of Takamatsu, Japan, to produce what would be the largest single window in the world: fifty-five feet long, fifteen feet high, and thirteen inches thick. It was shipped over from Japan in five sections. Upon arrival in Monterey, each section was lifted from the truck, lowered into the as-yet-roofless building, and laid flat on the bottom of the tank. The Nippura technicians constructed an insulated house over all five sections; they then slowly and uniformly heated the panels to the correct temperature for the chemical bonding that would join them into one massive window. At the end of this critical process, the window was slowly cooled to room temperature.

The Bigge Company, specialists in moving large and fragile objects, was hired to perform the delicate task of lifting and moving the giant window into its place in the concrete wall on one side of the tank. First their giant crane, which was parked on Cannery Row, tipped up the thirty-eight-ton piece of plastic from a horizontal to a vertical position. Then very slowly, an inch at a time, it was lifted and moved forward until it was in place, resting on neoprene pads in the concrete sill. Now the specialists from Nippura took over and injected silicone sealant around the window's edge. After a week of cure time, the tank was filled. Not a drop leaked from the window. The whole operation was an impressive achievement.

Creating the three-dimensional curved back of the tank was a major challenge for the engineers. Since a smooth, three-dimensional curve, like the inside of an egg, was certainly not practical to form in concrete, the solution they came up with was to pour the structural concrete in straight sections and then fit a curved fiberglass liner in front of the concrete. Water would be on both sides of this liner, so the pressure would be equal on both sides of the fiberglass. The liner only needed to be strong enough to support itself and to withstand any seismic forces generated during an earthquake.

Over the years I've had considerable experience with aquarium tank coatings of various types, and I have yet to come across one that holds up for any length of time. Sooner or later—usually sooner—the coatings all fail. The blue urethane on the inside of the Steinhart Aquarium's Fish Roundabout started to peel off the wall after just two years. The back wall of Sea World's Shark Encounter was painted with black epoxy that quite effectively made the back wall disappear from view when the tank was brand new. However, stress cracks soon developed in the concrete, and these needed to be patched and repainted. The paint over the patches changed color when wet, causing vertical light-colored stripes that were quite visible.

I knew that the exhibit we were building in Monterey would be home to fish so large and so fragile that they could never be safely moved. Draining and repainting the tank was therefore out of the question. We needed a blue finish that would remain in perfect condition for the lifetime of the tank itself.

The solution, it seemed, was colored glass tile, such as that used in Greek and Roman times to make mosaics, some of which are still in existence today after thousands of years. Glass tile never fades, and it is totally impervious to seawater. If applied with epoxy and sealed with an epoxy grout, it would likely stay in place for many, many years. Even if individual tiles should come loose, they could easily be glued back into place with a nontoxic epoxy that would cure underwater. With this coating, barring a catastrophic failure of some kind, like a major crack in the concrete tank itself, there should never be a need to move any of the animals to work on the tank.

A second and vitally important benefit of the tile surface was that although the individual tiles couldn't be seen by visitors thirty to fifty feet away, the fish would see them when they swam close. Living in a world of blue, tuna and probably most open ocean fishes are believed to have excellent vision in the blue range of the spectrum, and we hoped that they would easily see—and avoid—the pattern of the blue tiles. An absolutely evenly colored blue wall with no visual irregularities could be very dangerous to these fast-swimming fish. It was critical for them to know just where the wall was.

A number of years ago, at the offshore island of Socorro in Mexico, I had taken underwater photographs that captured the beautiful blue color of open ocean water. We sent these photographs to a tile manufacturer, Mosaicos Venecianos de México in Cuernavaca, to see how closely they could match the different shades of blue. The samples they sent back were very close to the colors of the open ocean. A test panel was made up with nine shades, from light blue at the top to deep blue at the bottom. We hung this panel in the Monterey Bay Habitats exhibit and looked at it through thirty feet of water. It looked great, and the breaks where one shade of blue changed to the next weren't visible at all. The tiles, we were confident, would work just fine in the Outer Bay exhibit, whose curved back wall would be forty to fifty-five feet away from viewers. We decided to place our order: 1,336,000 tiles, each one three-quarters of an inch square.

The tiles were delivered, and the Kreysler Company staff began the installation. Because the smooth, sloped liner inside the tank made it impractical to set up scaffolding, the tile was applied from two floating rafts. The installers worked from one end of the tank to the other, laying tile in three-foot-wide bands. As each level was finished, more water was added to the tank to float the rafts a little higher.

About three-quarters of the tile had been applied when we decided to fill the tank all the way and take a look at it. To our horror, we saw that even through fifty feet of water the different bands of blue were quite visible. What had looked great on our small-scale test panel did not work at full scale.

This was an awful development. We tried staggering the tiles where one color met the next to see if the bands would disappear, but with no luck: they were almost as obvious. The banded appearance of the back wall was simply unacceptable. We decided that a single, uniform blue was our only option. It was certainly better than horizontal stripes.

This correction was going to be an expensive hit on the budget, and it would also have an impact on the scheduling of the conditioning of the filters and the fish collecting, acclimation, and moving that all needed to be done before opening day. We would still be able to meet the already publicized opening date, but it would be tight, with no margin for any more delays.

Marty Manson, project manager for the exhibits in the new wing, her friend "Slim," and I flew to Mexico. With Slim, a professional race car driver, at the wheel, we made it from Mexico City to the tile company in Cuernavaca, and there we selected the best blue color with which to redo the entire back wall of the tank. The new batch of tile was produced on time, trucked to Monterey, and installed. When the tank was filled, the curved back wall looked good, with no corners or angles visible to spoil the illusion of endless water. Up close, the individual tiles were visible enough to warn the tuna that there was a wall. I felt confident that many years from now it would still look as good; there would be no reason to have to remove fish to repair a failing wall finish. However, as with anything totally new, only time would tell.

MACKEREL AND MORE

The large pelagic-fish exhibit would be the centerpiece of the Outer Bay Wing, but there were many smaller exhibits to design as well. The principal interpretive themes of the new wing were the two main types of open ocean life: the plankton, those animals and plants that drift with the ocean currents, such as jellies; and the nekton, animals that actively swim, including fishes, squids, and marine mammals.

We'd already done a popular special exhibit on jellies. We also knew that other aquarium staff had seen it and were planning their own jelly displays. To stay at the forefront of the friendly competition among public aquariums, we knew that we had to create something truly dramatic with our new jelly exhibits.

John Christiansen took on the challenge. Expanding on the tank design we developed for "Planet of the Jellies," he created exhibits that were to be quite a challenge to fabricate but would prove to be awesome in both their size and their visual impact. The largest, and one that rivals the million-gallon open ocean tank in its magnificent beauty, is a giant kreisel for sea nettles, twenty-two feet long and ten feet high. Shimmering comb jellies, delicate egg-yolk jellies, and elegant purple-striped jellies add to the quiet wonder of this gallery.

In keeping with the theme of swimmers, John created a twenty-foot-diameter overhead ring of endlessly swimming anchovies. This dramatic living sculpture, the first exhibit the visitor encounters in the new building, set the mood of the Outer Bay Wing.

The mackerel is an important pelagic species in both the Pacific and Atlantic Oceans, and we knew we had to have these fish on exhibit. However, although they're not difficult to collect and they adapt quickly to aquarium life, our past experiences with them had not been good.

The problem with mackerel, as with many species of the open sea, where food is scarce and patchy in distribution, is that they're genetically programmed to eat any time they see food. It's as if they don't know when they might get their next meal—even if they're on a regular feeding schedule in an aquarium. In the past, we had displayed mackerel in a multispecies tank. In such a situation, feeding the animals becomes a problem, especially making sure that the slower and more deliberate eaters get their fair share. With a school of speedy mackerel patrolling the upper levels, much of the food never makes it to the bottom. Meanwhile, the mackerel overeat and become grossly fat, such that they almost waddle through the water rather than swimming through it. But we couldn't put them on a diet without starving the slower fish near the bottom.

The solution was to create an exhibit where the mackerel were all by themselves. This way we could feed them only what they needed for metabolism and growth and they would keep their sleek, streamlined figures.

With the jellies we'd been able to simulate the limitless blue of endless water. To achieve that same impression with a school of constantly swimming mackerel, John and I came up with a tank in the shape of a racetrack or, thinking in three dimensions, an elongated bagel. Inside the oval hole of the bagel, lights would be suspended, illuminating a translucent blue panel set just across the track from the visitor's viewing window. As the fish swam by the window, they would be silhouetted against the blue background. The window itself would be small compared to the size of the tank, keeping the sides, top, and bottom out of sight. The effect was just what we had achieved with the jellies: a tank that looked as if it wasn't there. As the fish swam around the racetrack, it was as if there was an endless stream of mackerel, appearing from nowhere, swimming across the blue of a limitless sea, and disappearing into nowhere. It was eye-catching, and we felt that with the right diet the fish should stay healthy and slim.

While the husbandry staff worked on the live displays, Don Hughes and the exhibits team were busy designing the gallery spaces of the Outer Bay Wing. In keeping with the fluid feeling of the ocean, they created curving walls and a wavy suspended ceiling of blue slats. Designer Jim Stahl invented a lighting device, installed above the ceiling, that created rippling patterns of light playing on the blue carpet below. It was a masterful touch, and one that added tremendously to the watery feeling of the space.

22

A MILLION-GALLON FISHBOWL

Our R&D efforts with molas, bonito, and barracuda having gone well, we were finally ready to tackle the major challenge of the new wing: to see if we could catch, transport, and keep tuna. To date, tuna hadn't been kept in any aquarium in the United States, though Tokyo Sea Life Park, which opened in 1989, had a large, successful exhibit for the sole purpose of displaying yellowfin, bluefin, and skipjack tuna.

Tuna are highly prized in Japan and are served raw as maguro sashimi and sushi throughout the country. Bluefin are especially sought after; a single large fish of prime quality—which might be flown in fresh from as far away as the Mediterranean, South Australia, or New England—will sell in Tokyo's Tsukiji Fish Market for tens of thousands of dollars. This incredible demand has led to a serious decline in the North Atlantic bluefin population. At the same time, however, extensive aquaculture operations have sprung up on Kyushu and Shikoku, the southernmost of the main islands of Japan, where, in protected bays, thousands of young bluefins are held in huge, anchored net pens until they grow to marketable size. The commercial aspect of raising tuna appeals, I believe, to anyone interested in aquariums, aquaculture techniques, conservation, and economics. The more fish we can raise, the less pressure there will be on the wild populations.

TUNA IN JAPAN

In September 1990 Yoshitaka Abe, formerly of Ueno Aquarium and now director of Tokyo Sea Life Park, generously invited Chuck Farwell, John O'Sullivan, and me to join their chief collector, Hiroshi Sakurai, on a trip to the park's tuna-holding facilities in the little fishing village of Kasasa on Kyushu Island.

We stayed in a Japanese country inn, where I was introduced to traditional Japanese customs, which sadly are dying out as Japan becomes ever more westernized. The beautiful wooden building appeared to be meticulously constructed with hand tools. The accommodations, graciously administered by the staff, were strictly old-style Japanese, with no luxuries such as beds or chairs. I'm afraid you need to experience this lifestyle from childhood, because my old Western bones and joints complained loudly as I tried to sleep on a skinny futon spread out on the tatami-mat floor.

Bathing was done squatting on a tiny wooden stool, washing oneself from a small basin, and drying with a towel that was only slightly larger than the washcloths we use back home. Following this cleansing, we luxuriated in a communal Japanese hot bath.

The next day Sakurai-san and Sea Life Park collector Toshi Matsuyama took us out on a thirty-foot Japanese fishing boat to catch bluefin tuna. After an hour's run from the harbor we were in the clear, blue water that tuna like. With the boat still moving forward, the fisherman let out two trolling lines from the stern. Each one had a wooden board a few feet ahead of the lure, which carried a small barbless hook. The two boards skimmed along on top of the water.

Very soon we had the first strike. The jerk of the fish hitting the lure flipped the board, which plunged broadside beneath the water. The increased drag set the hook and let the fisherman know he had a fish. Hand over hand he pulled the line in, lifted a little ten-inch bluefin tuna on board, and quickly dropped the fish into a plastic bucket of seawater. The bucket had a taut nylon line tied across the top, which

Catching two bluefin tuna out of the rearing pen. (Photo courtesy John O'Sullivan)

caught the hook and neatly flipped it out of the tuna's mouth, releasing the fish. It was a pretty slick operation. The fisherman then carried the bucket to the front of the boat and gently released the tuna into one of several small holding tanks below deck level.

I was surprised to see that each holding tank was square in shape, not round, since I'd always assumed that a tank without corners would be easier for a fish to navigate. The fish, however, immediately started to swim around and very neatly turned before it even came close to running into the sides of the tank. A number of small bluefin were caught that day; I was very impressed at the efficiency of the operation and how well the fish behaved in their tanks.

At the end of the trip the tuna were carefully transferred one by one in buckets to one of the large pens anchored in the small bay. Here the fish would be fed daily and in a year's time would grow large enough to transport to Sea Life Park.

Leaving the little fishing village, we set off for the fishing port of Kashikajima on the island of Shikoku. Here were more large tuna-hold-

Transferring bluefin tuna into the ship's tanks for the two-day voyage to Tokyo Sea Life Park. (Photo courtesy John O'Sullivan)

ing net pens tied up alongside a narrow floating dock. While we watched, a seagoing ship about two hundred feet in length pulled into the harbor and tied up next to one of the tuna pens. It was one of several ships built specifically to carry live fish to the Tokyo fish market, but this time the Tokyo aquarium was chartering it to transport tuna for their exhibit.

We were about to see how tuna were transferred from a grow-out pen to the holding tanks of the ship. The ship's boom was swung out over the floating dock; suspended from its steel cable was a large, smooth, nylon-reinforced bag in the shape of a deep bucket. The water-filled bag hung just above the floating walkway next to the tuna pen between two fishermen. Each man held a stout pole fitted with a short nylon line and barbless lure. Slapping the pole and lure on the water, the first fisherman instantly had a tuna hooked; in one smooth motion he hoisted the two-foot-long fish out of the water and into the suspended bag. The hook flipped out of the tuna's mouth at about the same instant the second fisherman swung another tuna over the bag. This split-second operation was accompanied by a great deal of splashing by the powerful tuna, and flying spray soaked everyone within ten feet. The heavy bag was then quickly hoisted up by the ship's crane and lowered into the first of five holding tanks. This procedure was repeated until ten fish had been transferred from the net pen to the holding tank.

The operation then shifted to the second holding tank, and so forth until, after a couple of hours, a total of fifty tuna—ten per tank—had been loaded on board. The holding tanks were quickly topped off with seawater from powerful pumps on board, then securely battened down to prevent sloshing in rough weather. In short order, the ship cast off and headed out to sea for the two-day run up to Tokyo.

We flew back to Tokyo, and two days later Hiroshi Sakurai drove us down to the docks before daylight to meet the ship arriving from Shikoku with its load of live tuna. Lined up on the dock were five huge flatbed truck-and-trailer rigs, each carrying two circular fiberglass holding tanks as wide as the beds of the trailers. These trucks would transport the fifty tuna to the aquarium, twenty miles away. There are docks that are much closer to the aquarium, but like the waters of most industrial cities around the world, Tokyo Bay is polluted. Because the ship pumps the tuna's life-sustaining water straight from the sea, this alternate port with clean seawater was the only choice.

To catch the fish from each of the ship's five deep holding tanks, the water level was lowered to knee depth and wet-suited aquarium staff climbed down into the tank. They used specially designed plastic stretchers that held a single fish plus enough seawater to support the fish. The stretcher-borne tuna were then lifted one by one out of the hold, carried down to the dock, and transferred into the first truck's holding tanks. As soon as ten fish had been loaded, that truck took off for the aquarium and the next one pulled into position for its load.

Arriving back at Tokyo Sea Life Park, we watched the unloading of the tuna from the truck tanks into the exhibit. It was a wet and wild operation, with pairs of aquarists literally running each tuna stretcher from the truck, into the building, and up two flights of stairs to release the fish. Most of the tuna swam away and joined the resident group without incident, but a couple took off like aquatic rockets and hit the transparent acrylic window, with fatal results. Two scuba divers were in the exhibit tank just in case a fish sank to the bottom and needed help getting up and going.

The entire tuna operation, from the initial capture in southern Japan, to the year spent feeding and raising them in net pens, to the ship and truck transport to Tokyo Sea Life Park, impressed on me the human effort and money that went into each tuna on display. I don't know how much each live tuna costs the aquarium, but it must be very expensive. We are most grateful to the generous and hospitable staff of Tokyo Sea Life Park for openly sharing the hard-earned knowledge they gained in working with tuna.

ELUSIVE ALBACORE

It was scary to think that in the course of planning the new wing, we had committed the Monterey Bay Aquarium to a million-gallon exhibit of which the key species would be tuna. We knew our aquarium couldn't afford anything like the investment Tokyo Sea Life Park made to get their tuna. It was imperative we come up with a way to do our collecting at considerably less cost.

The situation in California is quite different from that of Japan. On the positive side, the tuna found off the coast of California are much larger, so we wouldn't have to spend a year growing the fish up to an exhibitable size. On the unknown side was the question of whether we could catch these wild tuna and then transport them safely, so that they remained healthy and active. I felt confident, though. The trip to Japan had been a great education for us, especially seeing Tokyo Sea Life Park's success in transporting bluefin weighing as much as twenty pounds.

In a way, the Japanese were fortunate to have an abundance of very small tuna that are easy to handle without causing injury to the fish. Off the California coast, we see much larger albacore, yellowfin, and bluefin tuna, weighing five pounds to forty pounds or more. Even a ten-pound tuna, which is mostly solid muscle, can potentially injure itself during its powerful struggles. To survive the trauma of capture, a fish that size must be handled correctly—and at this point we didn't know what that way was, or even if there was a correct way to handle them. We had a lot to learn ahead of us.

Upon our return to Monterey we made preparations for our first serious venture into working with tuna. Every late summer and fall, schools of albacore tuna sweep through California waters on their migration around the North Pacific. If the water temperature along the coast is warm, they may come as close as twenty miles from Monterey.

Such was the case during the El Niño of 1983–84. We had gone out in our twenty-five-foot collecting boat, *Lucile,* and caught two albacore. It was an opportunity to find out if they could be transported in the modest-sized holding tank on our boat. The tank turned out to be too small and the fish didn't survive the two-hour trip back to the aquarium. However, we did learn that we needed a bigger boat with larger holding tanks.

In addition to a good-sized transport tank, we needed a large, long-term holding tank back at the aquarium. There simply was no space at the aquarium for a tank of this size, but our sister institution, MBARI, twenty miles away in Moss Landing, had an old fish-packing warehouse that they said we could use until the new exhibit wing was completed. Taking them up on their kind offer, we installed a thirty-eight-foot-diameter, six-foot-deep fiberglass tank with a life-support system of sand filters, aeration, and ozone treatment. Vertical stripes of black tape were put all around the inside of the tank to let the fish see that there was a wall there.

Obtaining seawater for the new forty-thousand-gallon holding tank was a bit of a problem. Applying for coastal permits to install intake and discharge pipes in the Bay would, we knew, be a time-consuming bureaucratic process, one we couldn't afford to embark on if we were going to meet our schedule. Fortunately, the building was located near the cooling water discharge pipe of the Moss Landing Pacific Gas and Electric power station; we therefore obtained permission to pull our seawater from their pipe and then to return our outflow water back into it. It wasn't ideal, but it worked.

One hurdle had been taken care of with our trip to Japan. As well as sharing their hard-earned knowledge of tuna husbandry with us, the staff at Tokyo Sea Life Park had graciously loaned us three of their custom-designed plastic tuna stretchers. Now, with the tank and life-support system finally completed, everything seemed ready to give live tuna a try.

In August 1991, we chartered Ted Dunn's long-range sportfishing boat, the *Shogun*, out of San Diego. One hundred ten feet in length, it carried three large holding tanks below deck; in addition, it had four smaller on-deck bait tanks for holding live anchovies. Ted told us that the previous year he'd caught a few small yellowfin tuna and kept them in one of his large holding tanks for a number of days, and they seemed to do just fine.

This was most encouraging. We were optimistic about keeping tuna alive long enough to bring them into Moss Landing Harbor. Here the *Shogun* could tie up at MBARI's dock across the narrow road from our holding tank in the old warehouse, and from there it would be a short, quick trip for each fish from the boat to our tank.

At last, the time came to board the boat in San Diego and begin our hunt for the elusive migrating albacore. Ted Dunn had heard through the fishermen's grapevine that albacore had been caught a hundred miles off San Diego. We headed there, but trolling feathered albacore lures for many hours produced nothing. He'd also heard that there were fish far to the north, outside the Farallon Islands off San Francisco. So we headed north, with two trolling lines out at all times. A day and a night's run got us to the Farallons, but although we spotted a couple of other boats, we hooked up no fish and heard no radio reports of others catching fish.

Continuing north, we unfortunately headed right into a progressively worsening swell coming down from a storm in the Gulf of Alaska. Finally, off Cape Mendocino, the swell was so large that Ted decided he'd better turn around and head back south. We were now running downhill with sixteen-foot swells chasing us, and there was danger of an extra large wave breaking right over the stern of the boat. With the trolling lines still out, Ted posted a crewmate to watch for such breaker swells. Whenever he saw a big one coming he gave a whistle blast, and Norm Kagawa, the skipper, would give the boat more power to keep ahead of it.

Finally, after many days of continuous fishing, we hooked up two nice fifteen-pound albacore on our way home. Hand over hand they were brought to the stern, lifted in by the leader, and dropped into the holding tank. They immediately started swimming slowly around the ten-foot-square tank without touching the walls. They were beautiful with their extended winglike pectoral fins, reminding me of penguins swimming underwater.

Almost at the end of our charter, we headed straight for Moss Landing Harbor, radioing ahead to the aquarium to let the husbandry staff know we were coming in with fish. They were waiting on the dock when we pulled in and tied up.

Draining the stern holding tank down to knee-deep level, John O'Sullivan and I donned our wet suits and climbed in the tank with the two fish. Gently guiding one albacore into the plastic stretcher, we lifted it and the supporting water up to the people on deck. The fish was then run across the road into the warehouse. The stretcher was lowered into the water of the holding tank and opened up, and the fish swam out. That was repeated with the second fish, and the two quickly joined up to form a tiny school of two. The fish looked great, but we were disappointed that we hadn't been able to find more to make a larger school.

Although the tuna seemed to be swimming well, during the next few days they showed no interest in eating. We tried several foods we knew albacore like, but nothing worked. Even delicious-looking live anchovies were ignored. We kept on trying, but they took no food and after six weeks they died. Their refusal to feed baffled us and we were at a loss about what to do about albacore in the future.

Our colleagues from Tokyo Sea Life Park visited us, and even though they hadn't worked with albacore, they said our tank was too shallow for tuna. During the next months we took their advice and added a large fiberglass ring to the top of the tank, raising its height from six to ten feet. We removed the black tape stripes in order to fiberglass the two sections together, then repainted the inside. The tank, filled with filtered and heated seawater, was once again ready for occupation. This time, we hoped things would go better.

John O'Sullivan, our industrious and also party-loving collector, decided we needed to celebrate the hard work the plumber, carpenters, fiberglassers, and painters of the Facilities Department had done to modify the tank. Snacks and several cases of beer were bought, and the party began. After a few beers, it was just a matter of time before the first person was tossed into the pool. Some of us had planned ahead and brought swimwear, but others weren't so fortunate: they went in clothes and all.

That summer we again chartered the *Shogun* and went after albacore. This trip was more successful and fifteen albacore were brought in.

We were optimistic that we could keep them alive this time. We suspected that, in addition to the tank being too shallow, the two albacore might have failed to feed because there were simply too few fish to form a school. Two fish do not make a school. Ken Norris, my former boss at Marineland and later a respected professor of marine biology at UC Santa Cruz, theorized that schooling fish require a minimum number to school, feel secure, and survive. The number varies somewhat with the size and species of fish, but the critical number is believed to be about six to ten individuals. We had fifteen—more than enough to form a healthy school.

As it turned out, we found we had more lessons to learn as a number of unexpected things went wrong. First, we realized we'd made a big mistake by failing to replace the black stripes. For some reason we'd thought that because the first two albacore showed no inclination to run into the tank wall, perhaps we didn't need them. That was definitely wrong. The new fish had a hard time seeing the uniformly colored wall when they were startled and took off at high speed. We lost several fish from hitting the wall. Another problem was with the electrical power. A split-second delay in the emergency lights coming on was enough to panic a fish, causing it to hit the tank wall. And finally, once again, we weren't able to encourage the fish to take any type of food, living or dead. We lost all the fish within a few weeks.

YELLOWFIN TUNA

Although albacore are the primary species of migrating tuna that come by Monterey, we decided that yellowfin tuna might be a more practical species to work with. Not only are they easier to find off San Diego, but we knew from Tokyo Sea Life Park's experiences that they do well in an aquarium. In the summer of 1993, then, Chuck Farwell and John O'Sullivan went on a collecting trip off San Diego; they caught twenty-seven yellowfin tuna and brought them on the *Shogun* up to Moss Landing. We were a little concerned about how the warm water–loving yellowfin would do when the boat was in the cooler water in northern California above Point Conception, but they made it.

The fish were unloaded and transferred one by one into the newly restriped tank. Unlike the albacore, the first yellowfin started to feed in just a few days, and by the end of the week they were all eating.

This was exciting to see after the albacore's discouraging lack of response. The behavior of the yellowfin made the decision for us: we would focus our energies on yellowfin and put off albacore until after the new wing had opened.

The following year two more trips were made and we moved some fish into the twenty-foot-diameter holding tank we had in our quarantine area at the aquarium.

Having our primary tuna tank located twenty miles from the aquarium meant we spent a lot of unproductive time driving back and forth. Not only that, but it soon became clear to us that collecting tuna off San Diego and running the boat three hundred miles at fifteen knots all the way up to Monterey to unload the fish was not only time-consuming but unnecessarily expensive. A more practical and economical solution, we decided, was to build a large transport unit that could be loaded on a tractor-trailer rig and driven from San Diego to Monterey in ten hours instead of two days.

The concept of transporting fishes long distances over land wasn't new; many years earlier, Marineland of the Pacific had such a tank that they used to haul fishes from Mexico to Los Angeles and also to transport threadfin shad (*Dorosoma petenense*) from California to Hawaii by ship. While I was curator at Sea World I had chartered that Marineland tank to drive Sea of Cortez fishes from Loreto, Baja California, to San Diego. More recently, Sea World had built a similar tank to transport sharks across the country. From the 1930s up until the 1960s, too, the Shedd Aquarium had used a specially constructed railroad car to transport fishes from Florida and California to their aquarium in Chicago.

John O'Sullivan, Chuck Farwell, and I designed a three-thousand-gallon tank equipped with its own pumping, aeration, and filtration system. For both strength and temperature insulation, it would be constructed by the Kreysler Company of urethane foam sandwiched between two layers of fiberglass. For a ten-hour trip, we thought, the well-insulated tank wouldn't need refrigeration, not even in the hundred-degree heat of the Central Valley. Like our other holding tanks, the inside would have black vertical stripes so the fish would see the tank's boundaries. Its life-support system would be on a separate pallet connected by hoses.

The plan was to build the tank and its separate life-support unit so that it could be lifted onto a large rented truck whenever we needed to move fish. (When full of water, the tank and its life-support pallet weighed thirty thousand pounds.) A professional truck driver would handle all the driving, so we'd be free to concentrate on the welfare of the fish and the functioning of the life-support system.

TUNA RESEARCH FACILITY

Stanford University's Hopkins Marine Station, next door to the aquarium, had just hired tuna physiologist Dr. Barbara Block. Encouraged by our successes with tuna, she was anxious to have a facility at Hopkins where she could keep tuna for her behavioral and physiological work. Together, Chuck Farwell, John Christiansen, and Barbara designed a state-of-the-art tuna holding and research facility that would meet the needs of both our institutions as well as satisfy the requirements of the tuna.

The Tuna Research and Conservation Center, financed by the aquarium and located on Hopkins property, opened in August 1995. It consists of three large holding tanks: two are thirty feet in diameter and six feet deep, holding thirty thousand gallons each, and the third is forty feet in diameter and ten feet deep, with a capacity of ninety thousand gallons. The design of the life-support system for each tank is based on lessons we'd learned from the water quality and animal load on the system at Moss Landing. The new seawater systems were improved with higher circulation and filtration rates, plus heating, aeration, and foam fractionation to remove fish wastes. This meant that a substantial population of tuna could be housed while still maintaining the high-quality water required by these open ocean fish.

When everything was finally completed—the permit process, the tank and life support, and the building to house it all—another celebration was held by the Hopkins and aquarium staffs. Unlike the highly informal party we threw at Moss Landing, this one was respectable and nobody ended up in a tank.

Now that we had a first-class holding facility, the next project was to move all the tuna from Moss Landing to the new tanks at Hopkins. A diesel truck and driver were chartered and the new tuna transport tank was loaded on. The tank was filled with warm, sixty-eight-degree water and driven over to the Moss Landing tuna facility. There the holding tank was drained down, and Chuck Farwell, John O'Sullivan, Carolyn Darrow, and aquarist Scotty Greenwald climbed down a ladder into the knee-deep water. Using a crowder made of smooth plastic, they separated two fish from the school of nervous tuna and guided them with their hands into the fish stretchers. These were lifted up and passed to others who ran them outside to the waiting truck.

Safely released into the tank, the two tuna circled without touching the tank wall. We were very relieved to see them adjust so well. Eighteen fish were loaded into the truck tank and driven over to Monterey. A second run was then made for the rest. This was the first time we'd moved fish this large, and all of them made it just fine.

BRINGING TUNA HOME

Our initial experimental work now complete, we felt confident that we could catch, transport, and keep small- to medium-sized yellowfin tuna. We also had three large holding tanks ready and waiting to stockpile fish for the new exhibit tank. Despite these successes, however, we still had a lot to learn about tuna husbandry.

When we first began working with the tuna we assumed that because they're warm-blooded fish with a high metabolism and rapid growth rate, they needed lots of high-energy food. We therefore fed them to satiation two times a day with herring, anchovies, and squid, all of which they ate eagerly.

One day we lost one of the larger, older fish when it collided with the wall. During the autopsy we realized that something was seriously wrong with this fish when globules of oil oozed out of the muscle tissue. This didn't look like any healthy sashimi we'd ever seen or would want to eat. The fish was grossly obese. A week later another tuna keeled over and died right in the middle of a feeding frenzy. This fish, which also had oil in its tissues, apparently died of a heart attack from being too fat.

It became obvious that we were feeding our fish way too much food, as well as food that was too high in fat. Chuck Farwell began a thorough investigation into the metabolism of different types of fish and the caloric composition of the different kinds of food fish. His findings allowed us to make some changes.

The first step was to put the fish on a strict diet of low-fat squid and smelt in place of the high-energy herring and anchovies, to try to reduce the fat they had accumulated. They also went from twice-daily feedings to a single feeding three times a week. Sometime later we lost another fish through an accident, and the autopsy revealed this fish was in much better shape. The tissues looked normal and weren't oozing oil.

While these problems were being sorted out, Chuck Farwell and John O'Sullivan came up with a scheme that would give us the seventy to eighty tuna we needed to open the new exhibit wing. In preparation for the following summer's collection, John set up a fifteen-foot-diameter holding tank on the dock of the Scripps Institution of Oceanography's Marine Ship Facility in San Diego Bay. The plan (which optimistically assumed a successful trip) was to catch forty-five tuna with the *Shogun,* fifteen of which would be loaded directly into the truck transport tank and driven to Monterey, while the thirty remaining fish would be housed temporarily in the holding tank on the Scripps dock. After the first batch was unloaded, the truck and transport tank—which by now had been christened the "Tunabago"—would return for the second batch, and finally the third. Meanwhile, the *Shogun* would head back out to sea for a second collection.

It was an ambitious plan, and if the two collections and six truck trips went well, we could conceivably end up with ninety tuna, giving us some extras to keep in reserve. The tuna, plus the bonito, barracuda, and molas, would be a fine collection of fish for the aquarium to open with.

The weather and the fishing were good. Although we didn't catch

The "Tunabago" bringing a load of tuna to the Monterey Bay Aquarium. (Photo courtesy John O'Sullivan)

the theoretical ninety fish, we ended up with seventy tuna, including the ones from previous trips, that were ready to move over to the new tank.

By now, some of the tuna from the Moss Landing tank had been with us for two years and were a lot larger that anything we'd ever attempted to move. We were a bit nervous about just how we would move all those large fish one at a time out of their holding tank, into the Tunabago, and then up to the top of the open ocean tank. The first part was straightforward; it was getting the fish from the truck three floors up to the exhibit that had us analyzing various possible scenarios—some of which were considerably more reasonable than others. Any method we used, we knew, would require that a fish be confined in a stretcher with a limited amount of oxygenated water for a least several minutes. Speed was of the essence.

First we considered the Japanese method of two people running each fish in its stretcher up the three flights to the tank. The second method was to use the freight elevator. A third option was to lift each tuna with the crane located on top of the building. Timing each method with a stopwatch, we found that the freight elevator was the slowest, the crane the second slowest, and running the fastest. However, we rejected the running method as aquarist abuse; it was simply much too exhausting.

So we decided to give the crane a try, and a date to carry out the transfer was scheduled. The operation had to be done at night, when traffic on Cannery Row was minimal. The truck's tank was drained down, and two wet suit–clad aquarists climbed inside. The first fish was guided into a stretcher and passed up and hooked onto the crane. On top, two waiting divers took the stretcher, opened it, and let the fish swim out into the exhibit.

We had no idea how the fish would react when it saw that huge space. Would it take off like a rocket and slam into the wall, or into the invisible but equally solid acrylic window? With the latter possibility in mind, we had hung a series of ropes in front of the large window as a precaution so the tuna would have something to cue in on. Fortunately, our worries proved groundless. We all watched nervously as the first tuna swam out and began to cruise around the tank. It showed no indication of panic or of running into the window or walls. That was a tremendous relief. The next fish and all those that followed did the same as, one by one, they joined up to form a school.

THE DREAM COMES TRUE

After releasing the tuna, we introduced the mola, bonito, and barracuda. By the time we'd added the last fish, the exhibit was looking very good. It was a beautiful sight to watch the separate schools using all the space in the big tank.

The bonito seemed perfectly normal. Some were exhibiting their characteristic behavior of rolling over and swimming partly on their sides, something I'd seen in a school of wild bonito many years ago at Catalina Island. The motion causes light from above to reflect off their sides and flash silver. We can only guess as to the function of this behavior, but it may have something to do with keeping the school together.

TRAINED FISH

So far, everything seemed to be coming together, and while the tuna work was going on, other aquarists were working with molas. These fish have several attributes that made them a key species for our new exhibit. First, they are truly open ocean animals. In addition, their potentially huge size and their strange but charismatic appearance would make them an eye-catching sight.

The problem we faced with them is that they're slow. Speed isn't important in the wild, where they feed on jellyfish—which are even slower. However, in the large exhibit tank they would be with schools of speedy bonito, yellowfin tuna, and barracuda. Getting food to the molas could be a serious problem.

Molas may be physically slow, but we discovered during our early experimental work with them that they're quick learners: they figured out how to take food from our hands within only one or two days. A mola we'd had several years before had also responded whenever it saw the white feeding pole. If the ability to modify behavior and quickly adjust to a new situation is a sign of intelligence, molas must be the smartest fish around. We decided to take advantage of this intelligence and train our mola to come to us to get its food (we found one mola per exhibit to be best).

Aquarists Tim Cooke and Bonnie Grey made a plastic yellow star and attached it to a pole. Just before and during every feeding, the pole-mounted star was placed below the water surface of the holding tank. Within a few days the mola had learned to associate the star with feeding time and would scull right over to be fed. Positive reinforcement worked well, and when the fish was transferred to the big exhibit tank, feeding progressed without any problems at all.

The stunning sea nettle exhibit. (Photo © 2000 Monterey Bay Aquarium Foundation. All rights reserved.)

At first the slower-swimming barracuda were intimidated at feeding times by the fast, aggressive tuna and bonito, but they eventually learned an effective food-grabbing technique of their own and from then on had no trouble getting what they needed. As for the mola, it had been thoroughly trained to come to the side of its holding tank to be hand-fed. In its new home, it remembered its training well and adapted quickly to its new surroundings.

We were now counting the days until the grand opening and we could all feel a mounting tension. One last worry remained, for which we feared there might be no solution. Being aware of the unpredictable availability of jellies in the wild, we'd planned well in advance to raise our own sea nettles for the new huge kreisel tank. Yet the ones we were growing were still rather small. With time, of course, they would fill the tank, but we were disappointed that the display might be less than spectacular on opening day.

The jelly gods smiled down on us, and we were saved by a stroke of good fortune. That year there just happened to be a tremendous abundance of beautiful, big sea nettles all along the northern California and Oregon coast. Dave Wrobel, the aquarist responsible for the jelly exhibits, took the boat out and collected enough adult sea nettles to fill the new tank. When he was through, it looked awesome, with some fifty pulsing, flaming-orange jellies silhouetted against the wall of translucent blue. The original dream had become a reality.

23

A NEW DIRECTION

THE MONTEREY BAY AQUARIUM has set new standards for quality and visitor expectation in the public aquarium world. Diligent attention to detail, interesting and informative exhibits, a wide variety of education programs, active volunteer and member programs, and excellent customer service and facility maintenance placed this aquarium a step above similar institutions. Many in the profession see us as the leading aquarium in the United States, if not the world.

This was a fascinating facility to be associated with, but after the opening of the Outer Bay Wing in 1996, followed within just a few weeks by David Packard's death, I began to take a hard look at my future. I had been privileged to play a part in realizing the dream of the founders of the aquarium and its benefactors, David and Lucile Packard. Their vision has brought ocean awareness, beauty, and just plain fun to millions of people. In my decades of work in the aquarium industry, I had never seen anything come together quite like the Monterey Bay Aquarium did. It was a synergistic combination of place, situation, and talented, creative people who responded to the Packard example of excellence.

The Packards' commitment to quality was inspiring and compelling. They gave the aquarium designers the luxuries of time and money to develop and perfect exhibit ideas, to explore and experiment and do the best job we could. This dedication to excellence was felt by everyone involved in the project, from contractors to laborers, graphic designers to biologists—all of us. This is not to imply that there were not problems or differences of opinion along the way; there were plenty. But somehow they were resolved, and the end result was something of which we were all uniformly proud.

For me, though, it was time to move along, and at the end of 1997 I retired from the Monterey Bay Aquarium.

Since my boyhood days in England trying to figure out what my life's work would be, I've learned more than I thought possible, solving problems I couldn't even have imagined existed back then. But somehow, along the way, I learned a lesson that surprised me. My goal as an aquarist and biologist was to build exhibits, to collect and care for animals, and to create satisfying experiences for people of all ages. That part wasn't new to me. What was new was the keen appreciation I'd developed for visitors and what it means to them to come in contact with the creatures that inhabit the world of water.

We may have only one chance to turn a visitor's head toward conservation and conscience and away from complacency—a chance that is offered through our exhibits. For an exhibit and a visitor to make a connection, those of us designing displays and defining messages must be cautious.

Do we assume the visitor knows nothing and it's our obligation to explain as much as we can? At one time, I would have answered yes to that question. My original goal was to bring as much factual understanding as possible to the visitor, to describe every detail of a creature's existence, from feeding to spawning to its relationship with the other creatures in its environment. And I expected the visitor to absorb all this eagerly.

Now I see things quite differently. I've come to realize that perhaps our true goal in the aquarium world is to inspire awe, to create a sense of wonder and appreciation that will grow into caring. Communicating facts is all well and good, but without awakening a sense of caring we have accomplished little.

Conservation has become a watchword in the aquarium industry, and for good reason. As the evidence accumulates, it is becoming clear that the ocean is highly vulnerable, its living resources finite, not infinite as many have thought. Oil spills and pollution are a constant threat, and overfishing has depleted most of the commercial fish populations beyond sustainable levels. Obviously, the most vital message we can pass on to the coming generation is the importance of protecting the oceans and their inhabitants. The future of humankind and countless other species depends upon it.

I believe the key role of the Monterey Bay Aquarium is to promote conservation, as it should be for every public aquarium. Certainly, it's the message I take with me into the world. Like all biologists, I'm concerned about the accelerating worldwide destruction of habitats, depletion of rain forests in Southeast Asia and North and South America, overgrazing of African lands, pollution of air and water, and the ongoing extinction of species. All of these are a direct result of pressures from out-of-control human population growth. Not only is the quality of life for the majority of people on our planet declining because of overpopulation, but an incredibly complex system that we do not understand is being affected, a system that is the life support for all creatures. Despite our remarkable technology, we humans are not above the physical and biological laws that govern this planet. If we are not able to use our brains and learn to control ourselves, something most unpleasant will control us. In the meantime, we're threatening opportunities for people of all ages to be exposed to the wonders of the natural world.

I can't do much about these issues alone. What I can do is encourage others to appreciate and care for the world of nature and all that lives in it. This is challenge enough in a world where many children—even some who live within a few miles of Monterey Bay—have never seen the ocean and perhaps never will. Aquariums offer many people, including those who know the ocean well from on top, the only opportunity they may ever have to see and appreciate the wonders of the life forms beneath the surface, an experience that, we can only hope, will motivate them to protect the varied habitats and creatures of that remarkable realm.

As exciting as the Monterey Bay Aquarium is, I am always reminded of one simple fact: the view from the aquarium decks looking out toward the Bay and beyond to the open ocean is, indeed, our best exhibit. The ocean waves, floating kelp beds, busy sea otters, seals and barking sea lions, the salty smell of the ocean air, and the underwater world are beyond compare. Bringing the natural environment closer to each person has been the motivation behind the adventures of my adult life.

As Ratty dreamily said about mariners in *The Wind in the Willows,* "Yes, it's the life, the only life, to live." That is definitely true for me.

BRANCHING OUT

Throughout my career I've received a number of invitations to consult on other aquarium projects, advising on exhibit design and animal collection techniques. I've taken advantage of several such requests, and in so doing I've traveled to many interesting places—and been paid for it!

Often, when such invitations come along, I feel I'm given more credit than I deserve, and expected to come up with more than I'm capable of offering. Some aquarium planners even seem to believe that, with my answers to their questions, they can build an aquarium as successful as Monterey, only in less time and for less money.

The most unusual opportunity began with an inquiry from the U.S. State Department asking if I would go to China to help design the first modern aquarium in that country. Although the Chinese would cover only travel and living expenses, I accepted because I was sure this would be the only chance I'd ever have to visit China. David Pittinger of the National Aquarium of Baltimore was also invited, and we were allowed to bring our wives, though we paid all their travel expenses.

The aquarium was to be located in the coastal city of Dalian in northeast China, near the border of North Korea. We had no advance information about their aquarium plans or even about the city. We received a travel itinerary and airline tickets on Chinese National Airlines, departing San Francisco on New Year's Day 1988. Because it was the middle of winter, we knew it would be cold in north China. Betty and I packed what cold-weather clothes we had, and, with our passports and some cash, we were off. After a long, four-movie flight we landed in Shanghai, cleared immigration, and flew on to Beijing, where we would recuperate from jet lag and relax a few days.

Our first sightseeing trip was to Tiananmen Square, where the following year hundreds of pro-democracy demonstrators were massacred. Pictures of the square are deceiving, not even beginning to suggest how big it really is. It can easily hold a million people. Within the square are the Museum of the Chinese Revolution, the Mao Zedong Mausoleum, and Tiananmen Gate with its fifty-foot-high portrait of Mao.

We got in line to see Mao's tomb and were immediately pulled aside by a soldier and told with much waving and pointing to go over to an adjacent building, where we were instructed to leave everything we were carrying. Nothing, including cameras, was allowed to be taken into the tomb. Armed soldiers quickly moved people past the glass enclosure that held Mao's embalmed body, allowing visitors only a few seconds' glance before they were hustled out the exit. This worship of a dead body is bizarre to me, but clearly Mao was still revered by a lot of Chinese.

The next day we visited the Beijing Zoo. China, of course, is the home of the giant panda, and the Beijing Zoo has donated a number of pandas to zoos around the world. I figured if any place would have an outstanding panda display, it would be here.

We took a taxi to the zoo, paid admission, and went in. The day before had been cold, and this day was even colder, and the zoo was bleak, the trees bare and very few animals visible. The panda exhibit consisted of a rather small cage with steel bars. Inside the cage were a water dish, a few pieces of bamboo, and two rather dirty pandas. I'd certainly expected something else. It was quite depressing.

On our last day in Beijing we went to see the Great Wall, one section of which is not far from the city. The Great Wall, the largest structure ever made by humans, was built as a defense against raids by the nomadic peoples of Mongolia. Originally constructed in several separate sections, the wall was completed in about 200 B.C. A thousand years later it was breached when Genghis Khan and his army conquered much of China. During the first 150 years of the Ming Dynasty (1368–1644), the wall was repaired and extended to a total length of

1,500 miles—half the width of North America, a fact that I found mind-boggling.

The massiveness of the wall—twenty feet thick at the base and about twenty-five-feet high—is impressive. Steps lead to the top, and from there you can walk for miles, seeing the wall as it stretches over the hills and mountains in the distance. The sight of the massive stones brought to mind the terrible suffering that must have gone into its construction. From what I've read, almost all of the workers were forced labor. People were captured, or "Shanghaied," and forced to work on the wall until they died.

The next day we flew from Beijing to Dalian to begin our aquarium business. We were met by an interpreter and a representative from the Dalian Museum of Natural History, who escorted us to our hotel and told us we would be guests at a formal Chinese banquet to be held that evening in our honor with the mayor and city officials of Dalian. It was quite an event, complete with toasts to friendship and success of the aquarium, accompanied by lots of smiling and nodding. I'm not an adventurous eater, and I found most of the food strange and unappetizing. The most memorable dish was the jellyfish salad, an assortment of lettuce with thin strips of jellyfish on top. The jellyfish had a firm, crunchy texture and, surprisingly, was quite good.

Early the next morning, Dave Pittinger and I were taken by car to the museum, while Betty and Twig Pittinger went sightseeing with an interpreter. The museum, which had distinctive Russian architecture, had been built during the Russian occupation of northeast China (1898–1904). Our meetings were held in the director's office, the only room in the entire museum, we noticed, that was heated. When schoolchildren, couples, and families came into the icy cold museum, they were bundled in thick down jackets and fur or wool hats.

The meeting began, and it quickly became apparent that we weren't there to help plan their aquarium. They'd already done that: they had complete drawings and a scale model of an aquarium designed by a Japanese firm. Rather, we were there to give our opinion of the Japanese proposal. I'm sure the Japanese company wasn't aware that the Chinese had brought us Americans in to critique their design. We felt that we'd been placed in quite an awkward position and I felt they'd been dishonest with both the Japanese and us.

The week finally ended with us submitting an evaluation of the proposed design, whereupon the museum staff expressed their gratitude for our help and informed us that another traditional banquet would be held for us that evening.

The next day we flew back to Beijing, stayed overnight, and then flew on to Hong Kong. There we visited the highly successful Ocean Park oceanarium, which Ken Norris had helped launch after he left Sea World in San Diego. As for the Dalian aquarium, the government later brought in a New Zealand firm to help with design and construction, and the aquarium was completed in 1995.

Another small aquarium project that I consulted on, in Peru just south of the capital of Lima, unfortunately wasn't completed, the victim of a skyrocketing inflation rate. But working with the design team was invigorating, and after my consultation was over Betty and I made a fascinating side trip to Machu Picchu, the lost city of the Incas high in the Andes.

Seeing Cuzco, at an elevation of 11,000 feet, from an Aeroperú Boeing 727 as it banked around a snow-covered mountain was quite something. Cuzco was the capital of the Inca Empire until the Spanish arrived in 1532 on their search for gold. Conquistador Francisco Pizarro captured Emperor Atahualpa, held him for gold ransom, and then executed him. With their leader gone and 50 percent of the population wiped out by smallpox brought by the Spaniards, the Inca civilization collapsed.

We arrived at our small hotel and, wanting to see and do as much as we could in the brief time we had, immediately arranged for a van to drive us to the nearby ruins of the Inca fortress Sacsayhuaman (pronounced "sexy woman"). Level ground is a rarity in the Andes, so our sightseeing involved a lot of climbing around the ruins. Every step was an effort, and we made frequent stops to catch our breath. In our eagerness to see it all, we failed to heed this warning.

Returning to town, I was hit with a splitting headache. I'd never experienced altitude sickness before, but clearly that's what I was suffering from. My body, used to the pressures at sea level or below, was not at all comfortable with this thin air so far above the ocean. Arriving at 11,000 feet and going hiking right away was a dumb thing to do; we should have taken it easy for the first day. The hotel staff gave us some tea brewed from the leaves of the coca plant, the local remedy for high altitude sickness. It didn't help. I took several aspirin and went to bed.

The next morning, fortunately, I felt pretty good, so we packed up and boarded the little train that runs from Cuzco to Machu Picchu. The Andes are so steep that the train climbs by means of a series of switchbacks. First it goes forward up a short section of straight track and stops. The engineer gets out and switches the track, and the train goes backward up the next section—and so on, back and forth, until it gets over the mountain. Then it follows the same procedure to go down the other side.

Finally the mountain train arrived at the little station at the foot of the steep mountain of Machu Picchu. The ancient Inca citadel sits 1,800 feet above the station and is surrounded by majestic peaks towering thousands of feet above. Built as a remote royal estate and religious retreat for the Inca kings, it was used up to the mid–sixteenth century and then abandoned, remaining undiscovered for 400 years until 1911, when historian and explorer Hiram Bingham learned of its existence from a local peasant.

We boarded a Chevy van that took us up the narrow dirt switchbacks to the top. On reaching the summit we looked down hundreds of feet, where we made out the rushing turbulence of the Urubamba River, carrying the melting snows of the Andes to join the Amazon. We were awestruck by the majesty of the setting and the mystery of the place as we tried to imagine the lives and activities of those who once lived here.

Dozens of stone buildings are mostly intact, lacking only their long-ago disintegrated wooden roofs. They are constructed of granite blocks, some weighing as much as a hundred tons, hand cut from natural outcrops and shaped so that they literally locked together and anchored to the mountain itself. The Incas' engineering and construction skills were truly amazing. Although they lived in one of the most geologically unstable regions of the world, the buildings still stand after hundreds of years and untold numbers of earthquakes.

Who were these people, and what was their story? Their complex, highly organized society spanned two thousand miles of what is now Peru and Bolivia, yet they had no written language or even the use of the wheel for travel. They had domesticated the llama and alpaca, using them for the transport of goods and for clothing materials. By means of terracing and irrigation, Inca engineers from centuries past had made a mountainous terrain fertile and productive. Surrounding the buildings at Machu Picchu, for example, were terraces cut into the steep mountainside, used to grow food such as the many indigenous varieties of potatoes and maize. Remarkable feats of construction were accomplished in such cities as Cuzco using wooden rollers to transport huge stone blocks, and a farflung network of stone-paved roads, bridges, and even ferries linked all parts of the empire.

The time came for us to take the windy van ride back down to the little station. We boarded the train and headed back to Cuzco, and eventually home to California, feeling subdued after experiencing such a perfect combination of natural and man-made grandeur. After a lifetime spent diving in the oceans, I felt great respect for the people who had overcome the monumental challenges they must have faced high in the clouds.

GOING HERE AND THERE

As I've waited in one line or another in some airport, either in this country or elsewhere, I have occasionally asked myself, "Why are you doing this?" The answer is easy. I enjoy sharing the value of aquariums with others, the creative process of developing new exhibit ideas, and the challenges associated with working with new animals.

The first major project I was asked to consult on was the Aquarium of the Americas in New Orleans in 1986. The primary exhibit was to be a large Gulf of Mexico display featuring an underwater view of an oil rig; there would also be a secondary Caribbean exhibit, as well as displays focusing on the fresh water of the Amazon and Mississippi River drainages. Working with five architectural firms was a bit of a challenge for all of us on the project and I learned that it's essential to hold fast to your beliefs and to stick to what you think will work.

In 1994 I was asked to consult with the Two Oceans Aquarium in Cape Town, South Africa. I jumped at the opportunity, in large part because it marked my first return to the country of my birth. Apart from helping with the aquarium design, I would be able to see the country now that the oppressive policy of apartheid had ended.

In many ways South Africa is physically similar to California. Both have a Mediterranean climate, with wet winters and hot, dry summers, and around the Cape of Good Hope there is upwelling of cool, nutrient-rich water, which, as in California, produces an abundant growth of marine algae. A key element in the plan for the new aquarium was a living kelp forest. We came up with a design incorporating the same parameters that make our Kelp Forest exhibit in Monterey so successful: exposure to the sun, a surge machine to produce water movement, and a continuous supply of cool, nutrient-rich water.

I spent a week in design meetings and was invited to spend a couple of days with Geoffrey Starke, the project director, and his wife at their beach house. The morning following my arrival the beach was littered with masses of many different varieties of kelp broken loose by the big waves of a midwinter storm. We went fishing in the kelp bed and caught a number of bluefish (*Pomatomus saltatrix*)—the very same species found on the Atlantic Coast of the United States. Geoff hooked and, after quite a battle, released a five-foot-long sevengill shark—the same species we had in the Monterey Bay Aquarium. This wasn't a shark we wanted to bring into Geoff's little fourteen-foot boat.

I enjoyed working with the talented, dedicated, and friendly people at the Two Oceans Aquarium. Betty and I returned to Cape Town in 1996 for its grand opening and took the opportunity to explore some of the country.

Consulting on the South Carolina Aquarium in Charleston proved a unique experience. One day in 1989 I got a call from architect Antoine Predock in New Mexico asking if I would participate with him and Richard Graef of Ace Design in Sausalito, California, in a design charette competition for the proposed aquarium. Richard and I had worked together in the past, and I was impressed with his work; I'd never heard of Antoine, but I would get to know and like him by the end of what turned out to be an intense four days.

The process involved five teams of architects and exhibit designers from around the country, all of whom gathered in Charleston. On Monday, at 8 A.M. sharp, we were given the instructions and materials for the charette; only then were we actually allowed to begin work. At the end of four days, prior to 5 P.M. on Thursday, we were to submit to City Hall a set of architectural drawings, a scale model of the aquarium building, and a complete exhibit package, with descriptions and species lists plus sketches and renderings of each of the exhibits.

We picked up our instructions, retired to a motel suite, and went to work. We spent the first day or so brainstorming ideas until we'd come up with rough exhibit concepts. Our plan would focus on the wealth of animal and plant life in the wetlands of South Carolina; in addition, it would have exhibits of the inshore and offshore habitats of the coast. After that, it was all nitty-gritty detail work. I designed the exhibits, Richard produced sketches and renderings, and the architectural team worked on the building plans and the model. We lived on coffee, soft drinks, and delivered fast food. We barely made the five o'clock deadline, but we did and we turned our creations in to City Hall. As intense and exciting as it had been, by now we were totally exhausted.

The entries were judged the next day, but no overall winner was selected. Instead the city decided to split the competition into categories. Our team won for best exhibit program, but another team won for building design. The city later selected those participants they felt did the best work and put together a working team to design the new South Carolina Aquarium.

Richard and I teamed up with Linda Rhodes, who had been project manager of the Monterey Bay Aquarium; exhibit designers Dick Lyons and Frank Zaremba; and Alan Eskew, the lead architect for the New Orleans Aquarium. Despite a number of unexpected delays, the South Carolina Aquarium finally opened on May 19, 2000, to an enthusiastic response.

Another interesting project, on which I worked in 1985 together with Linda Rhodes and exhibit designer Jim Peterson, was the proposed Hawaiian Ocean Center in Honolulu. The focus was to be an unusual blend of marine biology, geology, and cultural anthropology showing how the early Hawaiians incorporated nature into their lives and legends. Sadly, the project fell victim to Hawaiian politics and was scuttled by the incoming governor. Among the innovative exhibits we had planned was the first open ocean exhibit for yellowfin tuna in the United States. Alas, that milestone had to wait for the Monterey Bay Aquarium to bring to fruition.

An unusual challenge came when I was asked to participate in brainstorming meetings to determine what to do about the degradation of tanks and the building at the National Aquarium in Baltimore (only two years older than the Monterey Bay Aquarium) due to corrosion. This was a prime example of what can happen down the road when a project relies on the lowest bidders and there's a lack of diligent project management.

The quality of the concrete work on the big tanks was poor, and in some places saltwater had penetrated all the way through the walls. The seepage into the concrete had caused the reinforcing steel to expand and pop off chunks of concrete, a process called spalling. The electrical system, which had been installed in steel conduit and junction boxes, was also failing from the effects of corrosion. We concluded that there was no possible quick or inexpensive fix. In the end the aquarium was closed for over a year while the necessary repairs were made, at a cost of many millions of dollars.

At Sea World and Steinhart I'd seen first-hand the results of the corrosive nature of seawater; this experience at Baltimore made me doubly glad I'd pushed for corrosion-resistant materials in Monterey.

An ongoing consulting project in which I have a high degree of interest is the New England Aquarium in Boston, the first of the new generation of aquariums. In 1987, John Prescott, the aquarium director and a friend from my early days at Marineland, invited a team of us to participate in a three-day brainstorming session on developing a major expansion and relocation of the aquarium for New England. Waterfront property had become so valuable in Boston that the old harborside aquarium was in an ideal position to make a good financial deal. They could sell the present site for commercial development and build a brand-new aquarium away from downtown for only a little more money than they would get for their property. We came up with some dramatic exhibits, and considerable follow-up work was done. Shortly afterward, however, the bottom fell out of the Boston real estate market, the value of the existing site dropped, and the project was shelved.

When John Prescott retired, his successor, Jerry Schubel, called another brainstorming meeting to develop a plan for the expansion of the aquarium at its present site. This project continues, but sadly, John Prescott did not live to see the completion of his dream; he lost a hard-fought battle with cancer and died in July of 1998. Throughout his career he had a major influence on the high standards of public aquariums and marine education throughout the world. He will be greatly missed.

I've worked on other consulting projects over the years, some small and some large. Whether I accept work is determined both by the professionalism of the participants and by my sense that the proposed facility will make a worthwhile contribution to increasing public awareness of the world of water. Few things, in the end, are as exciting to me as the possibility of a new aquarium that will open the eyes of new audiences and inspire them to conserve the aquatic life of the streams, lakes, and oceans of this, the Water Planet.

THOUGHTS UPON LOOKING BACK

A number of times I've been asked to name the best or biggest aquarium in the world. Such questions are impossible to answer. What does "best" really mean—most entertaining? most educational? having the greatest variety of animals or the most accurate representations of real habitats? As for size, do you judge that by square footage, the number of exhibits, total volume of water, or annual attendance figures? The "best" is probably the aquarium that has the greatest influence on its audience, the one that stimulates visitors into action to protect our deteriorating environment—an influence that's impossible to confirm or measure.

There's no such thing as the "ideal" aquarium. Each aquarium has some excellent and some less-than-exciting features. Ultimately, the success of an aquarium is greatly influenced by its location, the skill and creativity of the architects, designers, and biologists, and, of course, the level of financial backing.

Those responsible for seeing an aquarium come into existence—developers, planners, financial backers, and city officials—all have their own reasons for seeing it built, whether it's to display nearby aquatic animals, to renew a rundown section of town, to educate the local population, or what have you. However, an aquarium isn't likely to make a worthwhile contribution if its sole purpose is to boost a lagging economy, or if it is an entertainment-only project that makes no attempt to satisfy the visitor's curiosity. Likewise, one with too strong an emphasis on education will have a difficult time attracting visitors. An aquarium must pique the visitor's interest before it can hope to get across its messages.

Although I am talking primarily about aquariums, much of what I say applies equally well to zoos, living museums, and giant wildlife parks that feature entertainment along with their live animals.

The need for balance is perhaps the most important element on a loose list of factors I've come up with that, in my opinion, any aquarium should incorporate to be successful. I believe aquarium visitors want, even crave, a connection with nature. Why else would they be there? On the one hand they look forward to seeing something familiar, but they also anticipate elements of surprise. The surprise may lie in some fascinating piece of information, but it may also be an exhibit so simple in design that it literally stops the visitor in his or her tracks—a tall fish tank with a shimmering school of silvery lookdowns, for example, or a stunning jelly exhibit.

After this balancing of the familiar and the surprising, education and entertainment, information and aesthetics, it is imperative that high standards in design and attention to detail be applied to each and every exhibit—especially if an attempt is being made to re-create a natural environment. Excellent examples in this regard can be seen at the Arizona-Sonora Desert Museum in Tucson. Another fine example is the Kelp Forest exhibit at the Monterey Bay Aquarium. Here people stop and stare in awe, often with a look of rapture on their faces, in response to the natural beauty of the creatures and the totality of the environment. The fact that they are looking into a man-made structure is almost completely forgotten. Artificial exhibits can be effective as well, but then it is even more important that care be taken to provide the fishes with an environment as close as possible to that from which they come. While there are notable exceptions, in general, the closer we get to realism, the more effective the exhibit. This includes eliminating visual cues like pipes, drains, or square, flat walls that might draw attention to the artificiality of the tank.

The use of interactive exhibits is another important design component contributing to an aquarium's success. Interactives such as microscopes, floating magnifiers, or flapper display cards, in which you lift a flap up to reveal an interesting new fact, all engage the visitor directly, thus facilitating the learning process. Providing opportunities in which visitors can come in direct contact with living creatures, both large and small, is invaluable, though such exhibits—touch pools being an especially popular and effective example—require great care in their design, staffing, and operation.

The most effective interaction, of course, is with a real-live staff member or a knowledgeable volunteer. These living resources are also the most expensive to maintain. Staff need to be paid, and volunteers require extensive training and ongoing support in the continually changing subject matter of an aquarium.

It's important, too, that the aquarium building, the grounds, and all the exhibits be well maintained. Litter ruins the aesthetic experience of the overall space, while dirty or scratched tanks only draw attention to the fact that what is being shown is not a natural environment.

Another issue that concerns me is institutional philosophy. Decisions made in the boardroom don't remain there; rather, the consequences of such decisions—whether faulty or wise—can be seen throughout the public display areas. This applies especially to animal care, which should be any organization's number-one priority. We are using animals in our exhibits to help get our messages across to our visitors. They are not just tools to be used to make money. I cannot emphasize enough the value of respect and care for living things. Institutions that are insensitive in their display of animals eventually pay the price of losing their validity, or their audiences; they also give the entire field a bad name. The real losers, though, are the animal victims.

Another key to creating a successful aquarium is a simple "bread-and-butter" management concept. A number of years ago aquarists were often somewhat derogatorily called "tank men," but now most aquarists, men and women, are knowledgeable biologists and scientists in their own right. Those who manage aquarists need to understand that they are working with professionals; when problems arise, they therefore should thoroughly investigate the ramifications of the situation and then follow up with sound decisions.

Finally, I have what I call "Dave's Rule" for aquarium design. Basically, this standard says that if an exhibit is easy to work on, it will be well maintained and will look good. And if it looks good, it will tell its story successfully. When little consideration has gone into how an aquarist does his or her job, it shows. If an exhibit or behind-the-scenes space is difficult to negotiate, the aquarist will have little motivation to take those extra steps that make a display aesthetically pleasing or the animals happy.

The world of aquariums is constantly changing. The early days of the fish-in-a-tank approach of the classic old aquariums gave way in the 1970s to 1990s to the highly successful mega-aquariums that incorporated extensive educational programs combined with impressive exhibits of natural environments. Their success has spawned a new type of aquarium whose primary purpose is entertainment, represented by the aquarium in the mall, the restaurant, or the gambling casino. These convey little or no information about the animals and their environment; their sole intent is visual impact. Although these exhibits do reach a new audience, I am concerned that they trivialize nature and do little to encourage the viewer to respect wildlife.

The popularity and success of the large aquariums in the United States and Japan has prompted a number of other countries to embark on aquarium projects of their own. Some of these have been highly successful, but regrettably, some have not. Far too often, having a world-class aquarium is seen as a symbol of status and is undertaken for political reasons rather than a desire to foster an understanding and appreciation for the natural world around us. Compounding the problem, the promoters frequently do not understand the complexity of maintaining a large aquarium, or the knowledge and skilled staff that are required to make it work.

As increasing human pressures continue to impact the natural environments of both the land and the world of water there will be a growing need for places like aquariums and zoos. Undoubtedly these facilities will evolve and change and there will be new state-of-the-art exhibits and new methods of communicating messages, but the ultimate purpose for their existence will not change: they will still be places where the wonderful creatures that share this planet with us can be seen and appreciated.

DINNER WITH JACQUES COUSTEAU

The world of public aquariums is a small one, and communication among them has generally been conducted by individuals rather than on an institutional level. To promote the larger exchange of information, however, in 1958 the prestigious Musée Océanographique et Aquarium de Monaco, founded by Prince Albert in 1910, organized the first International Congress of Aquariology. Although the conference was very successful, thirty long years passed before a second one was organized. It, too, would take place in Monaco.

I was fortunate to be able to present a paper at the 1988 conference, which I attended with three other staff members from Monterey. Betty accompanied me, and we had plans to visit my sister in England afterward. The meetings were excellent; it was thrilling to share my work and hear what other biologists were doing in aquariums all over the world.

A farewell banquet was held in the famous Monaco casino. When Betty and I arrived we were escorted to a table at the front of the huge banquet hall. Already seated were the director of the Acuario Nacional de Cuba, a scientist and his wife from the Monaco Aquarium, and our friends John and Sandy Prescott from the New England Aquarium—across from whom I immediately recognized Jacques Cousteau.

I was completely blown away as, for the first time, I fully realized the worldwide reputation that the Monterey Bay Aquarium had achieved. Because of it, I had an opportunity to dine with the man who in many ways was responsible for my being there that very day. I told Cousteau that his book *The Silent World* had inspired me to take up scuba diving in its infancy and to follow his example by exploring the world beneath the sea.

Our lives, however, had taken different paths. Whereas he had chosen to show people the undersea world through his films, I wanted to bring people into contact with the undersea world through living animals in aquariums. This difference stimulated some lively discussion of the most effective way of reaching people.

Even though he was on the board of the Monaco Aquarium, Jacques Cousteau had come to believe that it is wrong to keep animals in aquariums. He said people could be influenced just as effectively through film, video, and interactive museum-type exhibits, without using live animals at all. Still, he had no answer when I posed the very real situation about what to do with an orphaned sea otter pup, only a few weeks old, that washed up on the beach. Do you leave it there to die, or do you pick it up and nurture it? If it fails, for whatever reason, to develop survival skills, you have no ethical alternative but to give it a home in as good an aquarium as you can create. These little sea otters then provide a valuable education for the public, becoming ambassadors for their species.

Cousteau said that at that very moment his organization was in the process of constructing an "Aquarium without Fish" in Paris that would open to the public the next year. I countered with the argument that humans have an innate need for contact with living animals, a need that aquariums satisfy admirably. Aquariums *with* live animals, I emphasized, are a most effective way to reach and influence people.

I heard later that two years after the opening of the "Aquarium without Fish," it closed from lack of attendance. However, Cousteau's one unsuccessful venture has been more than offset by the many films he made, which have been seen by millions of people all over the world. He did a great deal to raise people's awareness of the oceans, and his legacy will continue for many years.

The conversation turned toward less controversial topics, and we ended up discussing how dolphins and whales "do it." Cousteau said he'd had some remarkably close encounters with cetaceans so preoccupied with the act of mating that they completely ignored his presence. He said that they overcome the difficulty of mating in a fluid medium by having an almost prehensile penis that acts as if it has an eye on the end of it and knows right where to go.

Our table was so engrossed in this humorous subject that we failed to notice that the several hundred other diners had all finished eating. Apparently etiquette didn't permit them to leave until Cousteau himself stood up, and when this was pointed out to him he chuckled and rose to his feet with a wave. We thanked our host for a wonderful conference and a most stimulating evening and made our farewells.

That evening I spent with Jacques Cousteau comes to mind whenever I think about my retirement. Like Cousteau's, my retirement years are simply a continuation of my life—though I like to think I might be a little wiser than I was when I started out so many years ago. When meeting young, eager marine biologists, I do my best to steer them in a direction that will give them as much satisfaction as I've had in my work. When they ask what they can do to get started, I pass along commonsense tips: Get an education. Learn to dive, so you can experience the underwater world firsthand. Volunteer if you can. Keep learning, diving, and working with the animals you love.

I will keep on doing what I've been doing. Whenever I can dive, I do. Whenever I can work with or talk about sea creatures, I do. Whenever someone has an intriguing proposal—a collecting trip, a new exhibit idea, a day's fishing—I respond. The satisfaction of being on the water with the sounds and smells of the salt air is part of my being.

In fact, it doesn't seem so long ago that I returned as a youngster from a day at the beach in Durban with my mother, father, and sister and slept, that night, with a fish under my pillow. No, not so long ago at all.

SELECTED READING

Allen, Gerald R., and D. Ross Robertson. 1994. *Fishes of the Tropical Eastern Pacific.* Honolulu: University of Hawaii Press.

Ellis, Richard, and John E. McCosker. 1991. *Great White Shark.* New York: Harper Collins.

Grahame, Kenneth. 1908. *The Wind in the Willows.* New York: Scribner.

Herald, Earl S. 1961. *Living Fishes of the World.* New York: Doubleday.

Klimley, A. Peter, and David G. Ainley, eds. 1996. *Great White Sharks: The Biology of* Carcharodon carcharias. New York: Academic Press.

Norris, Kenneth S. 1974. *The Porpoise Watcher.* New York: Norton.

Taylor, Leighton R. 1993. *Aquariums: Windows to Nature.* New York: Prentice-Hall.

Walton, Izaak, and Charles Cotton. [1676] 1989. *The Compleat Angler.* Oxford: Oxford University Press.

Weinberg, Samantha. 1999. *A Fish Caught in Time: The Search for the Coelacanth.* London: Fourth Estate.

INDEX

Italic page numbers denote illustrations.

Abalone jingle shell (*Pododesmus cepio*), 33
Abbott, Don, 185
Abe, Yoshitaka, 259, 285
Acania (research vessel), 197
Acipenser transmontanus (white sturgeon), 207
Adcock, Dick, 55
Adioryx suborbitalis (soldierfish), 76
Albacore, 269, 270, 289–93
Alexander, Ralph, 151
Alijos Rocks, 132
Allen, Neil, 211, 218, 220
Alligator gars, 114, 115
Alligators, 49–50
Allocentrotus fragilis (fragile sea urchin), 13
Ampullae of Lorenzini, 224
Anchoveta, 176
Anderson, Andy, 199
Anderson, Chris, 200
Anderson, David, 42–44, 45, 46, 47
Andes, the, 309–11
Anemone, 1–2, 33, 39, 50, 255
Angelfish, 170; of the Revillagigedos, 131, 134, *136;* of Sea of Cortez, 44–45, 96
Angelos, Chris, 242
Anguilla (eel), 224
Animal rights activists, 100–101
Anisotremus davidsonii (sargo), 19
Anthias spp. (basslets), 170–71
Anthomastus ritteri (mushroom soft coral), 267
Apogon retrosella (cardinalfish), 76
Aqua Lung regulator, 6, *7*, 8
Aquarists: goal/tasks of, 211–13, 215; management of, 317–18
Aquarium of the Americas (New Orleans), 311
Aquarium tanks: artificial rocks for, 194, 197–98, 200–201; biological filtration in, 128; cleaning/maintaining, 212–13; disease's threat to, 188–89, 212, 273–74; first whale display in, 18; fish unsuitable for, 53, 194; Kreisel type of, 259–60; natural rocks for, 34–35, 58–59, 98, 101–2; open sea design of, 273, 277–81; for pelagic sharks, 105; public's response to, 191–92, 262–63, 265; seawater's corrosion of, 314; for Sea World restaurant, 38–39; sensory system type of, 147–49; temperature-controlled, 273; as touch tanks, 99–101, 128, 129, 317. *See also* Holding tanks; Transport tanks

"Aquarium without Fish" (Paris), 320
Archerfish (*Toxotes jaculator*), 148–49
Arizona-Sonora Desert Museum (Tucson), 316
Armstrong, Jody, 236
Artificial rocks, 194, 197–98, 199–201
Atractoscion nobilis (white sea bass), 19
Aurelia aurita (moon jelly), 258, 259, *260*, 272

Bahía de la Concepción, 247
Bahía de Los Angeles, 248
Baja California: map of, *68*. *See also* Cabo San Lucas; Los Frailes; San Felipe; Sea of Cortez
Bandar, Ray, 55, 56
Bang sticks, 132–33, 135
Bankia spp. (shipworms), 26
Barracuda, 19, 270, 276, 301
Bass, 19, 39, 207, 226, 228, 255
Basslets (*Anthias* and *Pseudanthias* spp.), 170–71
Bates, Ken, 218, 222
Bat ray (*Myliobatis californica*), 19
Bats, fish-eating, 56
Baxter, Chuck, 184–85, 195
Baylis, Derek, 199
Beach, Tom, 131–32
Beijing Zoo, 307
Bell, Willard, 70, 87
Bends, the, 181
Beroe spp. (comb jellies), 261, 282
Between Pacific Tides (Ricketts), 52
Big Emma (sevengill shark), 222–24, *223*
Bignose shark (*Carcharhinus altimus*), 123
Blackeye goby (*Coryphopterus nicholsii*), 102
Black ghost knifefish (*Gnathonemus petersi*), 148
Blackie (shark-catching dog), 152–53, *153*, 156, 159
Black jack (*Caranx lugubris*), 139
Black sea bass (*Stereolepis gigas*), 19, 39
Blacksmith fish (*Chromis punctipinnis*), 225, 228
Blacktip reef shark (*Carcharhinus melanopterus*), 150, 151, 152–53, *153*, 155–56, *157*, 158–59
Blacktip shark (*Carcharhinus limbatus*), 133
Blind goby (*Typhlogobius californiensis*), 29
Block, Barbara, 295
Bluebanded goby (*Lythrypnus dalli*), 23
Bluefin tuna, 284, 285–89, *286*, *287*
Bluefish (*Pomatomus saltatrix*), 312
Blue shark (*Prionace glauca*), 40, 104, *104*, 217; captivity research on, 105–7, *106*, *107*, 108–9; eyesight of, 220; metabolic functions of, 109–10
Blue-spotted jawfish (*Opisthognathus rosenblatti*), 83
Boats: for collection of fish, 25, 39, 55, 225–26, 291; Powell's building of, 11–12; and tidal fluctuation, 116
Bocaccio rockfish (*Sebastes paucispinis*), 187, 255
Bonito (*Sarda chilensis*), 19, 270, 276, 299
Bonnethead shark (*Sphyrna tiburo*), 112
Borthwick, Don, 96–97
Boston Whaler, 90–91
Bottlenose dolphin (*Tursiops truncatus*), 16, 18
Breeden, John, 165, 168
Brocato, Frank, 18
Brooks, Ernie, 225
Broomtail grouper (*Mycteroperca xenarcha*), 20
Brorsen, Steve, 241
Brosmophysis marginata (red brotula), 53
Bull shark (*Carcharhinus leucas*), 103, 110–15, 220
Bumphead parrotfish (*Scarus perrico*), 74
Burgess, Kent, 67
Burghart, Glenn, 62–63
Burnett, Nancy, 184–85
Burnett, Robin, 184–85, 195

Cabo Pulmo, 92
Cabo San Lucas, *69*, 94–96, 143; land-based collecting at, 90–91; road/ferry trip to, 87–90, *90*; Sea World's expeditions to, 67, *68*, 69–72, *85*, 85–86; submarine canyon at, 69, 84–85; tales of sharks at, 70–71

Calandrino, Boots, 18
California Academy of Sciences, 49; marooned researchers of, 140–41; Sea of Cortez expedition of, 54–58. *See also* Steinhart Aquarium
California barracuda (*Sphyraena californica*), 19, 270, 276, 301
California moray eel (*Gymnothorax mordax*), 20
California scorpionfish (*Scorpaena guttata*), 20, 25
Callianassa affinis (ghost shrimp), 29
Cannery Row, 51, 52; aquarium project on, 184, 185–86
Canton Island (Kanton Island): history of, 150–51; shark collecting at, 151–54, *153*, 156–58
Cape Town (South Africa), 312
Caranx lugubris (black jack), 139
Carcharhinus: albimarginatus (silvertip shark), 134, 138, *139; altimus* (bignose shark), 123; *amblyrhynchos* (gray reef shark), 156–57, 182; *galapagensis* (Galápagos shark), 133, 137; *leucas* (bull shark), 103, 110–15, 220; *limbatus* (blacktip shark), 133; *melanopterus* (blacktip reef shark), 150, 151, 152–53, *153*, 155–56, *157*, 158–59; *milberti* (sandbar shark), 103; *obscurus* (dusky shark), 41. *See also* Sharks
Carcharodon carcharias. See Great white shark
Cardinalfish (*Apogon retrosella*), 76
Carlson, Bruce, 150, 151, 205
Carlton, John "Muggs," 133, 141
Carmel Bay, 54, 197–98
Carnival (Mazatlán), 89–90, 92
Cassiopea xamachana (upside-down jelly), 263
Casteñares, Larry, 133
Castro, Al, 149
Catalina goby (*Lythrypnus dalli*), 23
Catalina Island (Channel Islands), 23, 31–32, 40
Caulolatilus princeps (ocean whitefish), 226, 229–30
Cea Egaño, Alfredo, 175, 177, 178
Centrostephanus coronatum (long-spined sea urchin), 23
Cephaloscyllium ventriosum (swell shark), 104, 226, 227–28
Cerralvo Island, 79
Chaetodon falcifer (scythe butterflyfish), *84*, 84–85
Chambered nautilus (*Nautilus pompilius*), 205, 209, 210
Chile, 175–78
China, 306, 307–9
Chinook salmon (*Oncorhynchus tshawytscha*), 208
Chivers, Dana, 54
Chivers, Dusty, 54, 55, 58
Chivers, Lynn, 54
Christiansen, John, 241–42; jelly tanks of, 260, 261, 281–82; mackerel tank of, 282–83; and ocean sunfish, *271*, 273
Chromis punctipinnis (blacksmith fish), 225, 228
Chrysaora fuscescens (sea nettle), 262, 282, *301*, 302
Chuckawalla lizard (*Sauromalus varius*), 57
Ciliated protozoan (*Cryptocaryon irritans*), 188–89
Clarion angelfish (*Holacanthus clarionensis*), 131, 134, *136*
Cleaner wrasse (*Labroides dimidiatus*), 170
Clingfish (*Gobiesox meandricus*), 29
Clinocottus spp. (sculpins), 28–29
Coelacanth (*Latimeria chalumnae*), 150, 164–65, *164*, 173–74; air transport of, 172, 173; legendary background of, 163–64
Collection: of albacore, 290–93; with bang sticks, 132–33, 135, 157; of barracuda, 276; of bluefin, in Japan, 285–89, *286*, *287;* of blue sharks, 105–7, *106*, *107*, 108–9, *109;* of bonito, 276; of diurnal fish, 72–75; with fish decompression tanks, 36, 38, 39; of flashlight fish, 166–69; with four teams, 241–42; of garden eels, 80–81; of hagfish, 256; of horn and swell sharks, 227–28; of jawfish, 83, 247–48; of king salmon, 208–9; land-based method of, 90–91; Mexican permits for,

Collection *(continued)*
126–27, 240, 241, 246; of molas, 272–73; of nearshore sharks, 116–20, 122–24, 123–24; with nets, 71, 72, 73, 74–75; of nocturnal fish, 75–77; of Panamic green moray, 244–46; of parrotfishes, 74–75; of pelagic sharks, 105–7, *106, 107;* with quinaldine anesthetic, 76, 80, 81; of razorfish, 81–82; of rockfish, 35–36, 39–40; with ROV, 266, 267; scientific, for Scripps, 132–33; of sea anemones, 1–2; of sevengill sharks, 217–20, 221–22; with slurp guns, 23–24; from submersible, 256; without harming fish, 82–85, 212, 289–90; of yellowfin tuna, 293–95; of yellowtail surgeonfish, 77–78
Comb jellies (*Pleurobrachia* and *Beroe* spp.), 261, 282
Comoro Islands: coelacanth expedition to, 163–66, 172; flashlight fish of, 166–70
The Compleat Angler (Walton), 4, 82
Concholepas (loco mollusk), 178
Conservation, promotion of, 304–6
Cooke, Tim, 275, 300
Copper sulfate treatment, 188–89
Coquimbo (Chile), 175, 176, 177, 178
Coral: of Canton Island, 151; dead brain, of Rarotonga, 182–83; of Monterey canyon, 267; of Socorro Island, 140; of Tahiti, 181–82
Cortez angelfish (*Pomacanthus zonipectus*), 44–45, 96
Corynactis californica (strawberry sea anemone), 33
Coryphaena hippurus (mahi-mahi), 243
Coryphopterus nicholsii (blackeye goby), 102
Cousteau, Jacques, xiii–xiv, 5, 319–21
Cousteau-Gagnan regulator, 6, *7*, 8
Crabs (*Paralithodes* spp.), 13; red land, 140, 141; sheep, 10, 26–27
Crosshatch triggerfish (*Xantichthys mento*), 134
Cryptocaryon irritans (ciliated protozoan), 188–89
Cuttlefish (*Sepia officinalis*), 210
Cuzco (Peru), 309, 311
Dailey, Murray, 120, 121–22
Dalian Museum of Natural History (China), 308–9
Darrow, Carolyn, 241, 296
Darwin, Charles, 56
Davis, Chuck, 188
Decompression sickness, 181
Decompression tanks for fish, 36, 37, 38
Deep sea: defined, 265; Monterey canyon fishes of, 266–68
Deep sea sole (*Embassichthys bathybius*), 255
Delta submersible, 252, 253–57
DeMotte, Dave, 66
Dempster, Bob, 49
Dendraster excentricus (sand dollar), 193–94
Diodon holocanthus (porcupinefish), 71
Disease, fish, 188–89, 212, 273–74
Dive n' Surf shop (Redondo Beach), 30–31
Dives: and the bends, 181; buddy system on, 72, 84; at Cabo San Lucas submarine canyon, 69; at Canton Island, 157–58; in Chilean waters, 177–78; at Easter Island, 180; at Farnsworth Bank, 31–32; how to start, *73;* at night, 9–10, 50, 53, 72–74; and nitrogen narcosis, 31–32; Powell's first, 8, 9–10; at Rarotonga, 182–83; at San Nicolas Island, 126; at Tahiti, 181–82; at Tanner Bank, 125–26; warming up after, 214. *See also* Collection
Diving gear: and bang sticks, 132–33, 135, 157; dry suits, 7; floating air supply, 46; for Marineland's underwater shows, 19–20; wet suits, 30–31
Dolphins: at Marineland exhibits, 16, 18, *19*, 20; mating of, 320; sonar of, 17; at Steinhart Aquarium, 50, 61, *61;* tool-using skill of, 20
Dorosoma petenense (threadfin shad), 294
Dry suits, 7
Dunn, Ted, 291
Dusky shark (*Carcharhinus obscurus*), 41

Earle, Sylvia, 165, 166, 168, 170
Easter Island, 178–80; *moai* heads of, *179*
Ebert, Dave, 218

Ebert, Earl, 26, 177
Eelpouts (*Melanostigma pammelas* and *Lycodapus mandibularis*), 268
Eels: bite from, 113; California moray, 20; collection of, 80–81, 244–46; garden, 55, 79–81, *80;* honeycomb moray, 170, *171;* migration of, 224; Panamic green moray, *142, 245;* and razorfish, 82
Egg-yolk jellies, 282
Electric fishes, 148
Elephant seal, 127, 129
Embassichthys bathybius (deep sea sole), 255
Enhydra lutris (sea otter), 186, 195–97, *196,* 236, 320
Enoshima Aquarium (Japan), 263, 264
Ensrud, Phyllis, 50, 64
Enteroctopus dofleini (giant Pacific octopus), 99, 209
Epinephelus itajara (giant grouper), 55
Eptatretus stouti (hagfish), 255–56
Ernogrammus walkeri (masked prickleback), 53
Ervin, Bill, 106, 107
Escherick, Homsey, Dodge, and Davis (architectural firm), 188
Eschrichtius robustus (gray whale), 37
Eskew, Alan, 313
Eugomphodus taurus (sand tiger shark), 103
Evermann, Barton Warren, 49, 141
Evermann, Mt. (Socorro Island), 141

Farnsworth Bank (Catalina Island), 31–32
Farwell, Chuck, 205, 219, 230, 240, 241, 261, 285; tuna collecting by, 293–94; tuna holding tanks of, 294–96, 297
Fawcett, Jerry, 15
Ferguson, Mark, 205, 241, 259
Fernandez, Papa and Mama, 44
Ferrante, Peter, 264
Filetail catshark (*Parmaturus xaniurus*), 267
Finespotted jawfish (*Opistognathus punctatus*), 56, 247, *247*
Fish-eating bat (*Pizonyx vivesi*), 56
Five Bells (diesel yacht), 70, 71, *85,* 86, 91
Flashlight fish (*Photoblepharon steinitzi*), 166–70, *167,* 172–73
Flying Tigers, 112, 115
Follett, Bill, 53
Fragile sea urchin (*Allocentrotus fragilis*), 13
FRC (fiberglass-reinforced cement), 194, 197–200
Fricke, Hans, 173
Friedman, Leonard, 131
Friends' School (Saffron Walden, England), 4–5
Fusiliers, 157

Gage, Carl, 193
Galápagos shark (*Carcharhinus galapagensis*), 133, 137
Garden eel (*Taenioconger digueti*), 55, 79–81, *80*
Garibaldi (*Hypsipops rubicunda*), 35, 226–27
Gecarcinus planatus (red land crabs), 140, 141
Genetic drift, 56–57
Geronimo (collecting boat), 25
Ghost shrimp (*Callianassa affinis*), 29
Giant grouper (*Epinephelus itajara*), 55
Giant jawfish (*Opisthognathus rhomaleus*), 83
Giant kelp (*Macrocystis pyrifera*), 186, 192–93, 198
Giant Pacific octopus (*Enteroctopus dofleini*), 99, 209
Giddings, Al, 165, 166, 169
Gill nets, 71
Girella nigricans (opaleye), 28
Globicephala macrorhynchus (pilot whale), 18
Gnathanodon speciosus (golden jack), 243–44
Gnathonemus petersi (black ghost knifefish), 148
Goatfish (*Mulloidichthys dentatus*), 71
Gobies, 23, 29, 102
Gobiesox meandricus (clingfish), 29
Golden jack (*Gnathanodon speciosus*), 243–44
Goldsberry, Don, 22
Goldsmith, Jerry, 18, 21, 29–30, 34–35
Gomez, Lloyd, 50, 64
Gonzaga Bay: angelfish of, 44–45; road trips to, 42–44, 46

Goodall, Jane, 3
Gordon, Ian, 234
Gorgonians, 247
Gotshall, Dan, 38
Graef, Richard, 312, 313
Gray reef shark (*Carcharhinus amblyrhynchos*), 156–57, 182
Gray smoothhound (*Mustelus californicus*), 216
Gray whale (*Eschrichtius robustus*), 37
Great Wall of China, 307–8
Great white shark (*Carcharodon carcharias*), 41, 104, 108; attempts to revive, 159–62, *161;* failed captivity of, 217, 233–35; metabolism of, 232; in Monterey Bay Aquarium, 230–32, *231;* in San Felipe, 117–19
Green, Tom, 64–65
Green-spotted rockfish (*Sebastes chlorostictus*), 40
Greenwald, Scotty, 296
Grey, Bonnie, 300
Gribbles (*Limnoria* spp.), 26
Griffin, Ted, 22
Groupers, 20, 55
Gulf grouper (*Mycteroperca jordani*), 20
Gulf of California. *See* Sea of Cortez
Gunther, Les, 151–52
Gymnothorax: castaneus (Panamic green moray), *142,* 244–46, *245; favigineus* (honeycomb moray), 170, *171; mordax* (California moray eel), 20
Gymnotids (electric fishes), 148

Haderlie, Gene, 197
Hagfish (*Eptatretus stouti*), 255–56
Halfmoon (*Medialuna californiensis*), 225–26
Haliotis rufescens (red abalone), 177
Halley, Pete, 159
Hamilton, Randy, 211, 242, 247
Hamner, William, 259, 260–61
Hand nets, 72, 73, 74–75
Harbor seal (*Phoca vitulina*), 50
Hart, John, 32, *107,* 116, 120
Hatchetfishes, 268
Hawaiian Ocean Center (Honolulu), 313–14
Hearst Castle (San Simeon), 144
Hemisquilla ensigera (mantis shrimp), 24
Herald, Earl, 48, 49, 55, 58–59, 62, 63–64, 65; coelacanth tank of, 150; death of, 146; Roundabout project of, 146–47
Heterodontus francisci (horn shark), 103–4, 226, 227
Hewlett, Bill, 198
Hexabranchus sanguineus (nudibranch), 170
Hexanchus griseus (sixgill shark), 220, 224
Hobson, Ted, 92
Holacanthus clarionensis (Clarion angelfish), 131, 134, *136*
Holding tanks: for eels, 81; at Hopkins research facility, 295–96; at Monterey Bay Aquarium, 206–7; for pelagic sharks, 105; temperature component of, 70, 98; for tuna, 286–88, *287,* 290, 292; for wrasses, 77
Honeycomb moray (*Gymnothorax favigineus*), 170, *171*
Hopkins Marine Station (Stanford University), 184; tuna research facility of, 295–96
Horn shark (*Heterodontus francisci*), 103–4, 226, 227
Hughes, Don, 249, 283
Humboldt Bay, 218
Humpback whale, song of male, 135–37
Hymer, Julie, 197, 211
Hyperion Sewage Treatment Plant (Los Angeles), 13
Hypsipops rubicunda (garibaldi), 35, 226–27

Imperial angel (*Pomacanthus imperator*), 170
Incas, 309–11
Interactive exhibits: importance of, 317; touch tanks, 99–101, 128, 129
International Congress of Aquariology (1988), 319
Intertidal pools. *See* Tide pools
Isurus oxyrhynchus (mako shark), 104, 110, 123, 220

Jackmackerel (*Trachurus symmetricus*), 226
Jacobs, Jake, 19
Japan, 284–89
Jawfish, 56, 83, 247, *247*
Jellies: alien invasion of, 264–65; and molas, 270, 272; Monterey Bay Aquarium's exhibits of, 262–64, 277, 281–82, *301*, 301–2; moon, 258, *260;* purple-striped, 261–62, 282; served at dinner, 308; tank designs for, 259–61; two-stage life cycle of, 258–59

Kagawa, Norm, 291
Kanton Island. *See* Canton Island
Kato, Susumu, 70–71, 115
Kelp bass (*Paralabrax clathratus*), 19, 226, 228
Kelp Forest exhibit (Monterey Bay Aquarium), *214,* 312, 316; collecting fishes for, 225–30; as experiment, 192–93; filling the tank of, 213, 215; tank design of, 188, 193–94
Kelp paddies, 226, 312
Keyes, Ray "Chub," 125, 156
Killer whale (*Orcinus orca*), 22
King salmon (*Oncorhynchus tshawytscha*), 208
Kinmont, Jerry, 116, 120
Kiwala, Bob, 1, 125, 143; Baja California trips of, 77, 88, 92, *93,* 96–97, 241; Canton Island trip of, 156, 157; and eel cage, 245; Rarotonga home of, 206; rockfish line of, 39–40; scythe butterflyfish of, 84–85; on shark transport flight, 112, 113–14, 115; Socorro trip of, 132, 136, 139; starting a dive, *73;* at Steinhart Aquarium, 62, 64, 65
Klay, Gerrit, 112, 115, 218
Kreisel tanks, 259–60
Kreysler Company, 280, 295

Labroides dimidiatus (cleaner wrasse), 170
Lagenorhynchus obliquidens (Pacific white-sided dolphin), 16, 18, *19,* 50, *61*
Lagios, Michael, 165
Lanternfishes, 268
Larson, Ron, 262
Las Animas Island, 55
Latimer, Courtney, 163
Latimeria chalumnae (coelacanth), 150, 164–65, *164,* 173–74; air transport of, 172, 173; legendary background of, 163–64
Lawrence, Marcia, 190
Lawrence, Sir Guy, 190
Leaffish (*Taenianotus triacanthus*), 170
Lemon shark (*Negaprion brevirostris*), 103, 110–11, 112–15, *114*
Leopard shark (*Triakis semifasciata*), 19, 39, 103, 216
Limnoria spp. (gribbles), 26
Lindsay, George, 55, 64
Lingcod (*Ophiodon elongatus*), 186–87
Loco mollusk (*Concholepas*), 178
Loligo opalescens (squid), 209
Long-spined sea urchin (*Centrostephanus coronatum*), 23
Lookdown (*Selene brevoorti*), 84
Los Angeles Harbor, 8
Los Frailes: beaching technique at, 242–43; equipment needs at, 240–41, 242, *243;* road trip to, 92–94
Lovejoy, Richie, *51*
Loxorhynchus grandis (sheep crab), 10, 26–27
Lucile (collecting boat), 225, 226, 275, 276, 290
Lycodapus mandibularis (eelpout), 268
Lythrypnus dalli (bluebanded goby), 23

Machu Picchu (the Andes), 309, 310–11
Mackerel, 282–83
Macrocystis pyrifera (giant kelp), 186, 192–93, 198
Mahi-mahi (*Coryphaena hippurus*), 243
Mako shark (*Isurus oxyrhynchus*), 104, 110, 123, 220
Manson, Marty, 281
Manta brevirostris (manta ray), 134
Manta ray (*Manta brevirostris*), 134
Mantis shrimp (*Hemisquilla ensigera*), 24
Maoris, 183
Mao Zedong Mausoleum, 307

Marine Aquarium Society of Los Angeles, 14, 15
Marineland of the Pacific (Palos Verdes), 5, 36, 38, 270; decline of, 21–22; dolphin research at, 17; Jewel Tanks of, 16; marine life exhibits of, 18–20; natural rockwork at, 34–35; as oceanarium, 15; tide pool tank at, 29–30; transport tanks of, 20–21, 294; wharf piling project at, 33–34
Marine mammals: at Marineland, 16, 18; for Mexico City exposition, 127–28, 129; at Sea World, 67. *See also* Dolphins; Sea otter; Whales
Marine snow, 254
Marisla (dive boat), 55
Mark I and II decompression tanks, 36, 37, 38
Marquardt, Bruce, 55
Marquardt, Roy, 55, 144
Martin, Linda, 201
Masked prickleback (*Ernogrammus walkeri*), 53
Mastigias papua (polka-dotted jelly), 263–64
Matsuyama, Toshi, 285
Mazatlán: Carnival in, 89–90, 92; road trip to, 87–89
MBARI (Monterey Bay Aquarium Research Institute): deep-sea research by, 266–68; founding of, 250–51; tuna holding tank at, 290
McColloch, Kelly, 75, 77, *80*, 92, *93*, 240, 242
McConnaughey, Ron, 132, 241
McCosker, John, 168, 182, 207; coelacanth expedition of, 163, 165–66, 172, 173; and Coquimbo aquarium, 175; and great white shark, 160, *161*, 233–34; Roundabout exhibit of, 146–47
McCosker, Sandra, 165
McFarland, Bill, 18
Medialuna californiensis (halfmoon), 225–26
Megalodicopia nians (predatory tunicate), 267
Meistrell, Bill and Bob, 30–31
Melanostigma pammelas (eelpout), 268
Mexico City fisheries exposition, 127–30
Meyer, Jeffrey, 146
Migration: of albacore, 290; of eels, 224; of molas, 271–72; of pelagic fishes, 269–70; of sevengill shark "Emma," 223–24
Millay, George, 22, 66, 67, 215
Miller, Ed, 159, 160
Mills, Claudia, 262
Moai (giant carved heads), 179, *179*, 180
Mola mola (ocean sunfish), 226, 269, 270–75, *271*, 300
Mollet, Henry, 225
Monterey Bay: deep-sea research in, 251, 266–68; jellies in, 258; migration of albacore to, 290; migration of molas to, 271–72; Powell's night dives in, 50, 53; sea anemones of, 1–2; shale beds of, 204; submersible's dive in, 252–57; water temperature of, 189
Monterey Bay Aquarium: architect of, 188; artificial rock experiments of, 194, 197–98, 199–201; cephalopod exhibit of, 209–10; deep-sea research at, 265–68; grand opening of, 235–37; jelly exhibits at, 262–64, 277, 281–82, *301*, 301–2; jelly research at, 257–60, 261–62; and MBARI, 251; "Mexico's Secret Sea" exhibit of, 239, 249; origins of, 184–86; David Packard's role at, 186, 198–99, 303–4; sand dollar exhibit of, 193–94; Sea of Cortez expedition of, 239–44; sea otter pup of, 195–97, *196;* shale exhibit of, 204; staff members of, 205–6, 211; success of, 238, 250; volunteers at, 213–14; "Whalefest" exhibit of, 238–39. *See also* Kelp Forest exhibit; Monterey Bay Habitats exhibit; Outer Bay Wing
Monterey Bay Aquarium Research Institute. *See* MBARI
Monterey Bay Habitats exhibit: collecting sevengill sharks for, 217–20, 221–22; great white shark in, 230–32, *231;* molas in, 272–73; predator-prey compromise of, 187–88; wharf pilings for, 201–4, *203*
Monterey submarine canyon: deep-sea research in, 251, 266–68; oxygen-

minimum layer of, 267; submersible's dive in, 252–57
Moonfish (*Vomer declivifrons*), 84
Moon jelly (*Aurelia aurita*), 258, 259, *260,* 272
Moorish idol (*Zanclus canescens*), 70
Morays. *See* Eels
Mormyrids (electric fishes), 148
Morone saxatilis (striped bass), 207
Morro Bay Aquarium, 195
Moss Landing Marine Laboratory (California State University), 266
Mulloidichthys dentatus (goatfish), 71
Muñoz, Francisco, 55
Muñoz, Pedro, 57
Murphy, Kym: Baja California trip of, 88, 89, 90; on Mt. Evermann, 141; shark collecting trip of, 112, 113, 133, *138;* shark sighting by, 40–41
Mushroom soft coral (*Anthomastus ritteri*), 267
Mustelus californicus (gray smoothhound), 216
Mycteroperca: jordani (gulf grouper), 20; *xenarcha* (broomtail grouper), 20
Myliobatis californica (bat ray), 19
Myripristis occidentalis (red squirrelfish), 76

Napoleon wrasse, 157
National Aquarium (Baltimore), 314
Natural selection process, 56–57
Nautilus pompilius (chambered nautilus), 205, 209, 210
Navarro, Leo, 72, 74, 81
El Navegante (sportfisher), 131, *135,* 137
Negaprion brevirostris (lemon shark), 103, 110–11, 112–15, *114*
Neoclinus blanchardi (sarcastic fringehead), 33–34
Neushal, Mike, 193
New Caledonia, 210
New England Aquarium (Boston), 314–15
Newman, Murray, 165
Nicklin, Chuck, 125, 133, 165, 169, 241; shark footage of, 137–39
Night dives, 50, 53; Powell's first, 9–10; skills/techniques of, 72–74
Nippura Company (Takamatsu, Japan), 278
Nitrogen narcosis, 32
Norris, Ken, 15, 21, 45, 66, 67, 309; dolphin research by, 17; management style of, 16, 18; on schooling fish, 293
North, Wheeler, 193
Notorhynchus cepedianus. See Sevengill shark
Nudibranch (*Hexabranchus sanguineus*), 170
Nygren, Scott, 222, 230, 241–42

Ocean Park (Hong Kong), 21, 309
Ocean sunfish (*Mola mola*), 226, 269, 270–75, *271,* 300
Ocean whitefish (*Caulolatilus princeps*), 226, 229–30
Octopus, 10, 29, 99, 209
Octopus bimaculatus (two-spot octopus), 10, 29
Oilfish (*Ruvettus* spp.), 164
Oil pneumonia, 69
Oligocottus spp. (sculpins), 28–29
Oncorhynchus tshawytscha (king, or Chinook, salmon), 208
O'Neill Rock (Revillagigedo Islands), 138
Opaleye (*Girella nigricans*), 28
Open ocean fishes. *See* Pelagic fishes
Ophiodon elongatus (lingcod), 186–87
Opisthognathus: punctatus (finespotted jawfish), 56, 247, *247; rhomaleus* (giant jawfish), 83; *rosenblatti* (blue-spotted jawfish), 83
Orca (*Orcinus orca*), 22
Orr, Robert, 55, 56
O'Sullivan, John, 292; fish transport devices of, 274–75, 276; Sea of Cortez trip of, 240, 241, 246, 247, 285; tuna collecting by, 293–94; tuna holding tanks of, 294–96, 297
Outer Bay Wing (Monterey Bay Aquarium): gallery spaces of, 283; jelly exhibits of, 281–82, *301,* 301–2; mackerel exhibit of, 282–83; molas of, 275, 300; open ocean tank of, 277–81; R&D needs for, 251, 257; from Tunabago to, 298–99

Oxyjulis californica (señorita), 227
Oxylebius pictus (painted greenling), 50, 102, 126

Pachycerianthus fimbriatus (tube-dwelling anemone), 1–2
Pacific white-sided dolphin (*Lagenorhynchus obliquidens*), 16, 18, *19*, 50, *61*
Packard, David, 235; Aquarium role of, 186, 189, 198–99, 214, 236, 250, 303–4; death of, 303; founding of MBARI by, 250–51
Packard, Julie, 184, 185, 195, 196, 215, 236; R&D philosophy of, 251, 257
Packard, Lucile, 186, 206, 236, 250, 303–4
Painted greenling (*Oxylebius pictus*), 50, 102, 126
Panamic green moray (*Gymnothorax castaneus*), *142*, 244–46, *245*
Panga (fishing skiff), 119
Panulirus interruptus (spiny lobster), 9–10
Paralabrax clathratus (kelp bass), 19, 226, 228
Paralabrax nebulifer (sand bass), 19
Paralithodes spp. (crabs), 13
Parasites: mola, 273–74; shark, 120; treatment of, 188–89, 212
Parker, B. J. and Frank, 7
Parmaturus xaniurus (filetail catshark), 267
Parrish, Chris, 55
Parrotfishes, 74–75, 149, 158
Payne, Roger, 136
Pedicillaria (pinching organs), 57
Pelagia: colorata (purple-striped jelly), 261–62, 282; *noctiluca* (an East Coast jelly), 261
Pelagic fishes: creating tank for, 273, 277–81; migration habits of, 269–70, 271–72
Pelagic sharks, *104*, 104–10, *107*. *See also* Great white shark
Permits for collection, 126–27, 240, 241, 246
Petaluma (California), 264
Peterson, Jim, 193–94, 236, 313
Phillips, Roger, 211
Phoca vitulina (harbor seal), 50
Photoblepharon steinitzi (flashlight fish), 166–70, *167*, 172–73
Piddocks (*Zirphaea* spp.), 204
Pilot whale (*Globicephala macrorhynchus*), 18
Pinnacles (Carmel Bay), 197–98
Pinochet, Augusto, 75
Pittinger, David, 306, 308
Pittinger, Twig, 308
Pizonyx vivesi (fish-eating bat), 56
"Planet of the Jellies" (Monterey Bay Aquarium exhibit), 263–65, 277
Planktonkreisel tank design, 259–60
Pleurobrachia spp. (comb jellies), 261, 282
Pododesmus cepio (abalone jingle shell), 33
Point Lobos (research vessel), 252–53, 254, 257
Polka-dotted jelly (*Mastigias papua*), 263–64
Pomacanthus: imperator (imperial angel), 170; *zonipectus* (Cortez angelfish), 45, 96
Pomatomus saltatrix (bluefish), 312
Porcupinefish (*Diodon holocanthus*), 71
Portuguese Bend (Palos Verdes), 28–29
Powder puff sea anemone, 39
Powell, Amy (Powell's daughter), 47, 48, 54, *61*, 147
Powell, Betty (Powell's wife), 7, 54, 207, 214; marriage to Powell, 13; with octopus, *12*; and Powell's relocations, 48, 147, 190; and sea otter pup, *196*, 196–97; on trips with Powell, 47, 91, 306–8, 312, 319
Powell, David: aquarium consulting projects of, 306, 308–9, 311–15; aquarium goals/philosophy of, 303–6, 315–19; boat-building by, 11–12; at Carnival in Mazatlán, 89–90; childhood interests of, 2–4, *3*; and Cortez angelfish name, 44–45; and Jacques Cousteau, 319–21; on Easter Island, 178–80; education of, 4–5; first dives of, 8, 9–10; fish decompression tank of, 36; fish-selling business of, 23; floating air supply of, 46; home aquarium of, 13–14; memorable dives of, 55–56, 125–26, 177–78, 181–83; military service of, 5; Monterey Bay Aquarium exhibits of, 186–88, 192–94; natural rock

projects of, 34–35, 58–59, 98, 101–2; open sea tank design of, 277–81; pictures of, *3, 7, 93, 107, 142;* on public response to exhibits, 191–92, 262–63; on road trips to Baja, 42–44, 46, 87–88, 92–94, 96–98; sensory system exhibits of, 147–49; and shark attack, 64–65; shark captivity research by, 104–10, *107, 108,* 234–35; shark transport experiment of, 110–15; on submersible dive, 252–57; tide pool tanks of, 29–30, 99–101; visiting the Andes, 309–11; visiting China, 307–8; wave machine of, 60–61; wharf piling projects of, 33–34, 201–4, *203*
Powell, Eve (Powell's daughter), 13, 48, 54, *61,* 147
Powell, Norman (Powell's cousin), 5–8, 9–10, 11, 50
Powell, Wilkie (Powell's cousin), 50
Praziquantil (deworming drug), 274
Predatory tunicate (*Megalodicopia nians*), 267
Predock, Antoine, 312
Prescott, John, 21, 319; death of, 18, 315; dolphin research by, 17; fish decompression tank of, 36–37; and New England Aquarium, 314–15; and scorpionfish, 25
Prescott, Sandy, 319
Prionace glauca. See Blue shark
Prionurus punctatus (yellowtail surgeonfish), 77–78, *78*
Pristis perotteti (sawfish), 21
Pseudanthias spp. (basslets), 170–71
Punta Colorada (Baja California), 92–93
Purple-striped jelly (*Pelagia colorata*), 261–62, 282
Python, 62–63

Quinaldine (fish anesthetic), 76, 80, 81
Quinn, Pat, 197, 211

Racanelli, John, 265
Rainbow wrasse (*Thalassoma lucasanum*), 76–77
Rand, Judy, 236
Rapa Nui (Easter Island), 179–80
Rarotonga (Cook Islands), 182–83
Rattlesnake, rattleless, 56
Red abalone (*Haliotis rufescens*), 177
Red brotula (*Brosmophysis marginata*), 53
Red land crabs (*Gecarcinus planatus*), 140, 141
Redondo Beach pier pilings, 32–34
Red squirrelfish (*Myripristis occidentalis*), 76
Reptiles: on Santa Catalina Island, 56–57; at Steinhart Aquarium, 62–64
Revillagigedo Islands: Clarion angelfish of, 131; humpback whales of, 136–37; live fish collecting at, 134–36; return trip from, 141, 143; scientific collecting at, 132, 133; sharks of, 137–39, *139. See also* Socorro Island
Rhizoprionodon longurio (sharpnose shark), 123
Rhodes, Linda, 184, 196, 236, 313
Ricketts, Ed "Doc," 50, *51,* 51–52, 59, 92, 239, 249
Rincodon typus (whale shark), 153–54
Rock and Waterscape Company (Los Angeles), 194
Rock-boring clams (*Zirphaea* spp.), 204
Rockfish: for Monterey Bay Aquarium, 186, 187; problematic collection of, 35–36, 39–40; on submersible dive, 255; on Tanner Bank dive, 126
Rocks: artificial, 194, 197–98, 199–201; natural, 34–35, 58–59, 98, 101–2
Rosenblatt, Dick, 85, 133
Rosy rockfish (*Sebastes rosaceus*), 40, 126
Rotenone (ichthyocide), 132, 133
Roundabout tanks: with predators and prey, 186–87; at Steinhart Aquarium, 146–47, 160, *161,* 279
ROV (remotely operated vehicle), 251, 257, 266, 267
Ruvettus spp. (oilfish), 164
Rypticus spp. (soapfish), 76

Sakurai, Hiroshi, 285, 288
Salmon, 186, 187, 208–9
Salter, Dave, 252–57
San Benedicto Island, 134
San Blas, 21

San Clemente Island, 40
Sandbar shark (*Carcharhinus milberti*), 103
Sand bass (*Paralabrax nebulifer*), 19
Sand dollar (*Dendraster excentricus*), 193–94
Sand tiger shark (*Eugomphodus taurus*), 103
San Felipe: collecting sharks at, 116–19, *117, 118;* losing boat at, 120–22; Sea World expedition to, 96–98; shark pool at, 122–23, *123*
San Luis Gonzaga, road trip to, 42–44
San Nicolas Island, 126
San Simeon Aquarium plan, 144–45
Santa Catalina Island (Channel Islands), 23, 31–32, 40
Santa Catalina Island (Gulf of California), 56, 57
Santiago (Chile), 176
Sarcastic fringehead (*Neoclinus blanchardi*), 33–34
Sarda chilensis (bonito), 19, 270, 276, 299
Sardines, 186
Sargo (*Anisotremus davidsonii*), 19
Sauromalus varius (chuckawalla lizard), 57
Sawfish (*Pristis perotteti*), 21
Scalloped hammerhead shark (*Sphyrna lewini*), 55, 133
Scarus perrico (bumphead parrotfish), 74
Schubel, Jerry, 315
Scientific Expedition to the Sea of Cortez (June 1964), 54–58
Scorpaena guttata (California scorpionfish), 20, 25
Scott, Don and Julia, 90, 94
Scripps Institution of Oceanography, 132, 173, 297
Sculpins (*Clinocottus* spp. and *Oligocottus* spp.), 28–29
Scythe butterflyfish (*Chaetodon falcifer*), *84,* 84–85
Sea anemone, 1–2, 33, 39, 50, 255
Sea Life Park (Hawaii), 21
Sea lions, and molas, 271
Seals: elephant, 127, 129; harbor, 50
Sea nettle (*Chrysaora fuscescens*), 262, 282, *301,* 302
Sea of Cortez: angelfish in, 44–45; aquarium exhibits of, 239–40; collecting diurnal fish in, 72–74; collecting nocturnal fish in, 75–77; garden eels in, *79,* 79–81, *80;* jawfishes in, 83; land-based collecting in, 90–91; losing boat in, 120–21; map of, *68;* 1964 Scientific Expedition to, 54–58; Panamic green morays in, 244–46; parrotfish in, 74–75; permits for collecting in, 126–27, 240, 241, 246; Powell's first dive in, 55–56; razorfish in, 81–82; restaurant aquarium fish from, 38–39; scythe butterflyfish in, *84,* 84–85; sharks in, 115–19, 122–24, *123;* yellowtail surgeonfish in, 77–78, *78. See also* Cabo San Lucas; Los Frailes; San Felipe
The Sea of Cortez (Steinbeck and Ricketts), 52, 92, 239
Sea otter (*Enhydra lutris*), 186, 195–97, *196,* 236, 320
Sea perch, 186
Searama (Galveston, Texas), shark transport from, 110–15
Sea urchin, 13; as bait, 227, 229; long-spined, 23; short-spined, 57
Sea World (San Diego): Cabo San Lucas expeditions of, 67, *68,* 69–72, *85,* 85–86, 87–89; founders of, 66–67; great white shark at, 233; Los Frailes expedition of, 92–94; losing boat of, 120–22; marine mammal shows at, 67; natural rock aquariums of, 98, 101–2; Revillagigedos expedition of, 131–37, *135;* San Felipe expedition of, 96–98; Sea of Cortez exhibit of, 86, 87; Shamu at, 22; shark reproduction at, 158–59; shark research program at, 104–10, *107, 108;* shark transport to, 110–15, 155–56, 294; tank coatings at, 279; touch tanks at, 99–101
Sebastes: chlorostictus (green-spotted rockfish), 40; *constellatus* (starry rockfish), 40; *miniatus* (vermilion rockfish), 40, 255; *paucispinis* (bocaccio rockfish), 187, 255; *rosaceus* (rosy rockfish), 40, 126
Selene brevoorti (lookdown), 84
Semicossyphus pulcher (sheephead), 226, 229
Señorita (*Oxyjulis californica*), 227

Sepia officinalis (cuttlefish), 210
Sergestid shrimp, 255
Seri, Paul, 230
Seriola dorsalis (yellowtail), 226
Sevengill shark (*Notorhynchus cepedianus*), *223;* attack by, 64–65; collection/release of, 218–24; described, 217; with great white shark, 232; in South Africa, 312
Shale beds, 204
Shamu (killer whale), 22
Sharks: attack by, 64–65; captivity problems with, 104–5, 108–10, 217, 232–35; caught by dog, 152–53, *153;* collection of nearshore, 116–19; collection of pelagic, 105–7, *106, 107;* collection of sevengill, 218–20, 221–22; collection of swell, 227–28; with coral reef habitat, 150; defense against, 132–33, 135; metabolism of, 109–10, 225, 232; migration of, 223–24; nearshore species of, 103–4; pelagic species of, 104, *104, 107, 108;* reproduction of, at Sea World, 158–59; of the Revillagigedos, 131, 137–39, *139;* seasonal appearance of, 119–20, 133–34; sensory systems of, 220, 224; swimming with, 111, 153–54; tacos made with, 123–24; tales about, in Cabo, 70–71; transport of, by air, 111–15, 154–55; transport tanks for pelagic, 105–6, 107, *108, 109. See also* Great white shark
Sharpnose shark (*Rhizoprionodon longurio*), 123
Shedd, Milt, 39, 66, 70, 72
Shedd Aquarium (Chicago), 294
Sheep crab (*Loxorhynchus grandis*), 10, 26–27
Sheephead (*Semicossyphus pulcher*), 226, 229
Shelley, Mark, 229, 232
Sheppard Rock (Cabo San Lucas), 71
Shima Marineland (Japan), 146
Shimura, Kazuko, 264
Shipworms (*Teredo* and *Bankia* spp.), 26
Shogun (sportfishing boat), 291, 293, 294, 297
Short-spined sea urchin (*Toxopneustes rosacea*), 57
Shrimp: ghost, 29; mantis, 24; red, 178, 255; sergestid, 255
The Silent World (Cousteau), 5, 319
Silvertip shark (*Carcharhinus albimarginatus*), 134, 138, *139*
Sixgill shark (*Hexanchus griseus*), 220, 224
Slurp gun, 23–24
Smooth hammerhead shark (*Sphyrna zygaena*), 104
Snakes, 62–64
Soapfish (*Rypticus* spp.), 76
Socorro Island, *135;* coral of, 140; humpback whales of, 136–37; live fish collecting at, 134–36; location of, 131; marooned researchers on, 140–41; return trip from, 141, 143; sharks of, 71, 137–39
Soldierfish (*Adioryx suborbitalis*), 76
Sommer, Freya, 211, 264; jelly research by, 257–58, 259–60, 261–62; mola research by, 272–74
Soupfin shark, 218
South Carolina Aquarium (Charleston), 312–13
Sphyraena californica (California barracuda), 19, 270, 276, 301
Sphyrna: lewini (scalloped hammerhead shark), 55, 133; *tiburo* (bonnethead shark), 112; *zygaena* (smooth hammerhead shark), 104
Spiny lobster (*Panulirus interruptus*), 9–10
Squid (*Loligo opalescens*), 209
Stahl, Jim, 283
Stanford Research Institute, 186
Starke, Geoffrey, 312
Starry rockfish (*Sebastes constellatus*), 40
Steinbeck, John, 52, 92, 239, 249
Steinhart, Ignatz, 49
Steinhart Aquarium (San Francisco): coelacanth expedition of, 163–66, 172; dolphin exhibit of, 61, *61;* Fish Roundabout of, 146–47, 186, 279; flashlight fish exhibit of, 173; great white shark at, 232–34; natural rockwork at, 58–59; public's response to, 191–92; renovation of, 48, 49–50; sensory system exhibits of, 147–49; sevengills at, 218; shark attack

Steinhart Aquarium *(continued)*
at, 64–65; snakes at, 62–64; tide pool experiments at, 59–61
Stereolepis gigas (black sea bass), 19, 39
Strawberry sea anemone (*Corynactis californica*), 33
Striped bass (*Morone saxatilis*), 207
Strobilation of jellies, 258–59
Sturgeon, 207
Sulawesi (Indonesia), 163
Sullivan, Dennis, 37
Sunset wrasse (*Thalassoma lutescens*), 76
Sweeney, Jay, 120, 129
Swell shark (*Cephaloscyllium ventriosum*), 104, 226, 227–28
Swimbladders, decompression of, 35–36, 38, 39, 72
Szymczak, Ray, 92, 93

Taenianotus triacanthus (leaffish), 170
Taenioconger digueti (garden eel), 55, 79–81, *80*
Tahiti, 181–82
Tamm, Sid, 262
Tanner Bank, dive at, 125–26
Taylor, Leighton, 153
Tealia sea anemone (*Urticina piscivora*), 39, 50
Teredo spp. (shipworms), 26
Thalassoma: lucasanum (rainbow wrasse), 76–77; *lutescens* (sunset wrasse), 76
Thornyhead, 255
Threadfin shad (*Dorosoma petenense*), 294
Thunnus albacares (yellowfin tuna), 134, 293–94, 296–97, 314
Tiananmen Square, 307
Tiburoneros (shark fishermen), 123–24
Tide pools: experiments to create, 29–30, 59–61; fluctuation levels in, 116; marine life in, 28–29; Ricketts' study of, 52; at San Simeon, 144; touch tank exhibit of, 99–101
Tima formosa (Japanese species of jelly), 263
Toba Aquarium (Japan), 173
Tokyo Sea Life Park, 284; collection of bluefin by, 285–89, *286, 287;* plastic stretchers from, 290–91; tank advice from, 292
Tonga John (Seone Tau), 152, 156
Totoaba macdonaldi (totuava), 20
Totuava (*Totoaba macdonaldi*), 20
Touch tank exhibits, 99–101, 128, 129, 317
Toxopneustes rosacea (short-spined sea urchin), 57
Toxotes jaculator (archerfish), 148–49
Trachurus symmetricus (jackmackerel), 226
Transport tanks: air shipment of, 58, 112–15, 154–55; for bonito and barracuda, 276; for bull shark, 111–12; lowered into exhibits, 219, 231; Marineland's design of, 20–21; for pelagic sharks, 105–6, 107, *108, 109;* Styrofoam boxes as, 58, 156, *157,* 240; for yellowfin tuna, 294–95, 297–98, *298*
Triakis semifasciata (leopard shark), 19, 39, 103, 216
Tsegeletos, George, 55
Tsukiji Fish Market (Tokyo), 284
Tucker, Tom, 207
Tuna: collection of albacore, 290–93; collection of bluefin, 285–89; diet for, 296–97; holding tanks for, 286–88, *287,* 290, 292; Hopkins holding facility for, 295–96; Japanese demand for, 284–85; plastic stretchers for, 288; rearing pen for, *286;* schooling of, 293; yellowfin, 134, 293–94
Tunabago (transport tank), 297, *298,* 298–99
Tuna Research and Conservation Center (Hopkins Marine Station), 295–96
Tunicates, 156, 267
Turner, Chuck, 26
Tursiops truncatus (bottlenose dolphin), 16
Two Oceans Aquarium (Cape Town, South Africa), 312
Two-spot octopus (*Octopus bimaculatus*), 10, 29
Tyler, Bertha and Dean, 195
Typhlogobius californiensis (blind goby), 29

Underwater World (Sydney, Australia), 234
University of Mexico City, 126–27

University of North Chile, 175, 177
Upside-down jelly (*Cassiopea xamachana*), 263
Upton, Bruce, 211, 241–42
Urticina piscivora (Tealia sea anemone), 39, 50
U.S. Divers Company (Westwood Village, California), 6, 8

Vancouver Aquarium, 165
Van Dykhuizen, Gilbert, 252; Baja collection trip of, 242, 244, 247; deep-sea research by, 266, 267; and sevengill sharks, 218, 219, 222, 225
Van Wormer, Bob, 92
Vermilion rockfish (*Sebastes miniatus*), 40, 255
Villalobos, Alejandro, 55
Vomer declivifrons (moonfish), 84

Waikiki Aquarium, 150, 151, 154, 210
Walker, Boyd, 25, 45, 53, 185
Walton, Izaak, 4, 82
Ward, Peter, 210
Wave machine, 60–61
Webster, Steve, 184–85, 195, 239, 247
Weekley, Jamie, 210
Weekley, Michael, 1, 205, 210
Western Biological Lab (Monterey), 50
Western Flyer (fishing boat), 52, 239, 249
Wet suits, 7, 30–31
Whales: first aquarium display of, 18; mating of, 320; Powell's encounter with, 37; song of humpback, 135–37; transport of, 112
Whale shark (*Rincodon typus*), 153–54
Wharf pilings: for Marineland exhibits, 32–34; for Monterey Bay exhibits, 187–88, 201–4, *203*
White sea bass (*Atractoscion nobilis*), 19
White sturgeon (*Acipenser transmontanus*), 207
Whitman, Tom, 111, 112
Wiggins, Ira, 55
Wilder, Randy, 242, 247
Williams, Tom, 195, 197, 222, 274
Wilson, J. Walter, 23–24
World War II, 3, 5
Wrasses: cleaner, 170; holding tanks for, 77; Napoleon, 157; rainbow, 76–77; razorfish, 81–82; sex change in, 229; sheephead, 226, 229; sunset, 76
Wrobel, Dave, 241–42, 302

Xantichthys mento (crosshatch triggerfish), 134

Yellowfin tuna (*Thunnus albacares*), 134, 293–94, 296–97, 314
Yellowtail (*Seriola dorsalis*), 226
Yellowtail surgeonfish (*Prionurus punctatus*), 77–78, *78*
Yoklavich, Mary, 252

Zanclus canescens (moorish idol), 70
Zaremba, Frank, 313
Zirphaea spp. (rock-boring clams, or piddocks), 204
Zumwalt, Don, 18

Text:	12/14.5 Adobe Garamond
Display:	Perpetua and Adobe Garamond
Composition:	Integrated Composition Systems
Printing and binding:	Edwards Brothers, Inc.
Index:	Patricia Deminna